MILLIPED ZOOLOGY

HUSBANDRY AND DIVERSITY OF THE CLASS DIPLOPODA

Orin A. McMonigle

COACHWHIP PUBLICATIONS
GREENVILLE, OHIO

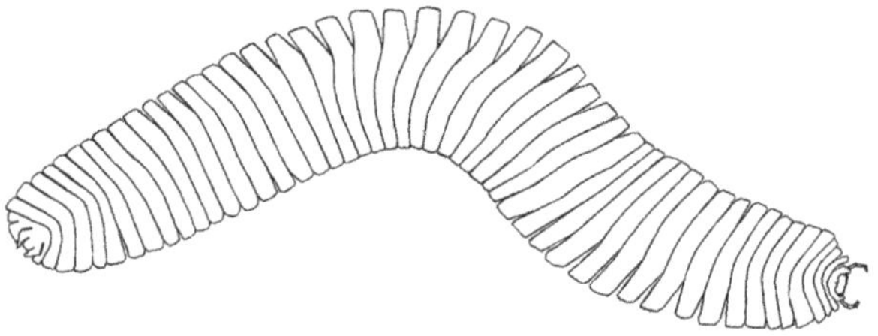

IN MEMORY OF ONDREJ HES

DEDICATED TO ROWLAND SHELLEY

Acknowledgments

I offer great appreciation to Annie Lastar for editorial assistance. I acknowledge Dr. Rowland Shelley, Dr. Richard Hoffman, Perry Adkison, Sam Floyd, Tom Larsen, Wade Harrell, Michael P. McNichols, and D. T. Anderson for content assistance in earlier texts. Special thanks is extended to God for the amazing beauty and diversity developed even in creatures of utility. I thank my loving wife Sylvia, daughters Kree and Gwynevere, and son Jax for sharing their time and space with these beautiful, utilitarian decomposers.

Coachwhip Publications // CoachwhipBooks.com

ISBN 1-61646-550-6
ISBN-13 978-1-61646-550-6

CONTENTS

(Order Spirostreptida) *Thyropygus* sp.; Thailand

(Order Spirostreptida) *Spirostreptus* sp., adult; Tanzania

INTRODUCTION

Milliped Zoology combines resources to provide a thorough and updated work on milliped husbandry and captive breeding efforts for select orders and species in the Class Diplopoda. Millipeds are well-established in captive stocks with a small but entrenched following of enthusiasts and zoologists around the globe. Despite sixty years of specimen availability for display and research, practical care and reproducible breeding information for the large-bodied spirostreptidans and spirobolidans only became available in 2005 with *Giant Millipedes: The Enthusiast's Handbook*. It was the first resource to provide dedicated enthusiasts and zoologists a tested methodology for breeding these animals and offered the concerned keeper dependable husbandry information. Still, the format was small with details limited to forty pages and fewer species.

Attempting to complete the literature, *Millipeds in Captivity* (2012) was a large-format, full-color expansion of *The Enthusiast's Handbook.* It contained information from various diplopod husbandry articles published 2002-2012 in *Invertebrates-Magazine*, Hoffman's toxifabricants, and other new chapters, while incorporating an additional seven years of rearing and longevity records. Although documentation of captive breeding and life cycle periods have remained surprisingly sparse, an additional eleven years of research has been included in this current work. *Milliped Zoology* is greatly expanded and updated from the 2012 text. This volume adds a new section on glomeridans along with details on how to breed them successfully, sections on platydesmidans and macrojulidans which have seen some attempts at captive husbandry in recent years, and data from thirteen additional *Invertebrates-Magazine* husbandry articles.

The term 'milliped' (rather than 'millipede') is used in most taxonomic and scientific works by American authors. In the 2005 husbandry handbook the only unused suggestion made by the late Dr. Richard L. Hoffman was for the text to employ the term 'milliped,' without the terminal *e*. 'Milliped' is used here in accordance with Hoffman (2000) and to honor a man whose extensive work on milliped taxonomy and biology may never be paralleled.

Milliped zoology has lost many workers since the first iteration of this work. Perry Adkison, 1963-2008, provided information and photos of an eastern milliped hunter and was the only reliable supplier of native *Apheloria*. I miss discussing millipeds with him. Dr. Hoffman, 1927-2012, a world-renowned myriapodologist, provided several specimen identifications, submitted the chapter on toxifabricants for the 2012 *Millipedes in Captivity* text, and provided editorial suggestions and text additions for the 2005 handbook. I will sincerely miss his answers to my questions. Dr. Rowland Shelley, 1942-2018, provided editorial comments and described two new species of *Floridobolus* as a direct result of identification assistance for the 2012 text. Barney Tomberlin, 1942-2021, was a singular source for western milliped hunters, *Orthoporus,* as well as some details for the earlier texts. I hope this greatly expanded book is a fitting tribute to these great men and their passion for myriapods.

Imports of exotic miliped species into the United States were no longer allowed after 2006 (shortly followed by Canada), so a number of species are only available to European diplopod enthusiasts. (Permitted imports are rare and other continents have not hosted an appreciable community of keepers.) The loss of common, inexpensive *Archispirostreptus gigas* as a starting point for the beginning zoologist is incalculable, but many large endemic species are spectacular arthropods. Over twenty years of displays at bug festivals the general public were commonly shocked at the size of *Narceus* spp., and responded with disbelief that such a large arthropod could be figuratively found in their back yard. Native and adventive U. S. species have taken the place of exotics in many local pet shops so

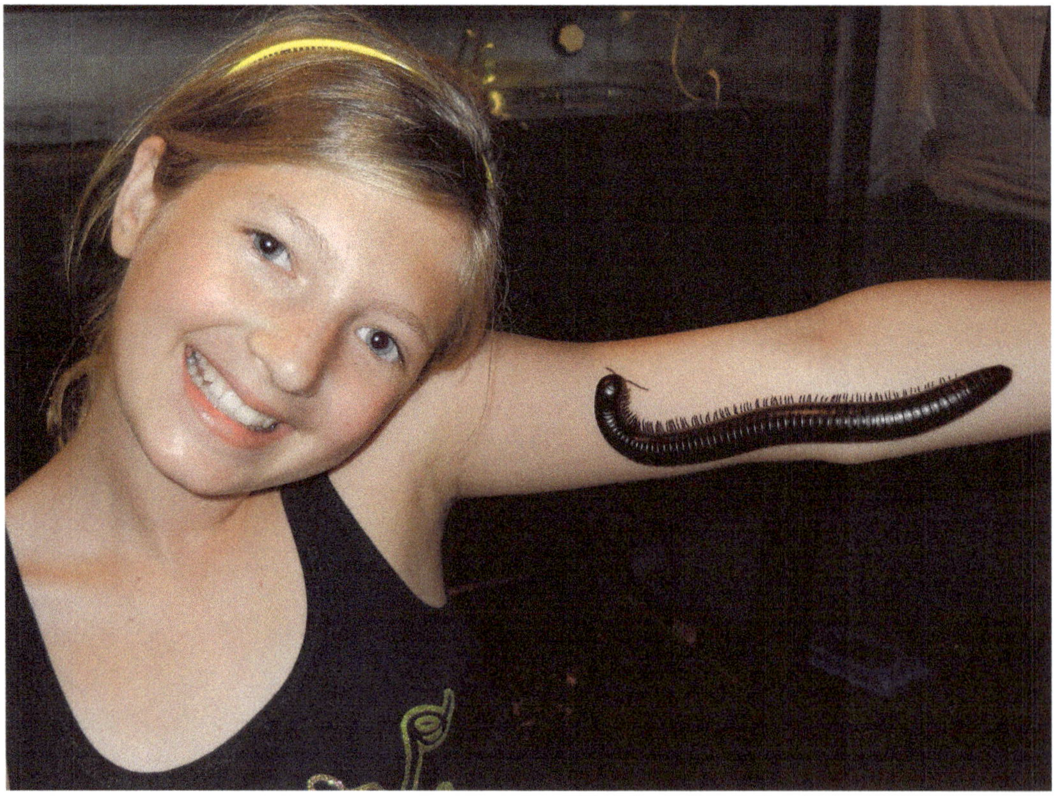

Gwynevere with captive-bred *Archispirostreptus gigas* (2012)

the availability and keeping of North American species has improved greatly. Specialty vendors intermittently carry a dozen different North American endemics today that were unheard of twenty years ago, although some, like *Apheloria virginiensis,* are more difficult to acquire.

Literature has improved over the decades, however resources for husbandry parameters useful to milliped zoologists are few. Checklists, as the name implies, lack information on biology, feeding, reproduction, and anatomical descriptions, and reference original and some subsequent accounts and distributional data. Some scientific papers over the past forty years delve into life histories to varying degrees, particularly regarding household pest species and almost never extend beyond F1 (filial or captive hatched) generations in the laboratory. Numerous terraria and small animal books contain a chapter on millipeds. Suggestions for care are commonly detrimental to the health of the animals and include lethal ventilation and recommended diets that would rapidly kill immatures and eventually starve adults. There are a few books, one a popular children's book, written specifically on millipeds and centipedes; the organisms possess some anatomical similarities, but the combination is forced and husbandry unrelated. Unfortunately, even the more dedicated resources were not restricted to diplopods and did not include firsthand husbandry and specific captive reproduction information prior to the 2005 *Giant Millipedes* handbook. In 2010 a book on pet millipeds by Sigling was translated from German and offered to the English-speaking hobby. The format is handsome and in full color with beautiful photos of many species not seen even in the heyday of milliped imports in the U.S. Unfortunately, very few North American species were included and what husbandry information is provided holds little utility, due partly to translation quality. In addition, diagrams copied from *The Biology of Millipedes* but not cited, and some glaring information errors distract from the value. *The Biology of Millipedes* (Hopkin and Read 1992) provides an overview of the class, but is based primarily on small-bodied European species, contains unfortunate errors (Shelley 1996), costs $150, and understandably lacks husbandry information.

Giant Millipedes: The Enthusiast's Handbook, **2005**

The details presented here are based on captive data and cannot reflect every possible captive situation nor are the suggested parameters meant to emulate natural conditions. Predators, scarcity of food, weather extremes, pests, and other natural conditions are avoided as much as possible. Since each female often produces thousands of eggs over a lifetime, if 99% of developing larvae were to die in a captive habitat, the survival rate would still greatly outpace natural replacement of parent quantity (two). Most natural conditions are avoided in captive culture to improve survival. The described husbandry suggestions and materials have been successful for many species over multiple generations. Methods may be modified with experience, but reinvention is more likely to result in failure than success. Time periods, lifespans, and supplemental foods are likely to vary from natural conditions and among specimens.

This text is meant to advance appreciation for millipeds and document successful methodology for maintaining consecutive generations in artificial habitats. Detailed information on captive habitat, substrate, food, rearing, and egg laying parameters are offered to inform zoologists interested in culturing this group of handsome creatures. Large-bodied millipeds have been a part of zoological displays since at least the 1960s (Stanek 1969; Walls 1999), but while captive breeding is more common than it was a decade ago, the animals seen for sale are mostly wild-caught. Since the early days, keepers have bred or at least collected eggs from wild females, but either did not understand what they did correctly or were unwilling to share the information. Even beginning zoologists and enthusiasts can use the instructions detailed here to keep their animals alive and healthy for years rather than months. Readers interested in breeding millipeds and willing to follow directions will discover many species are easily maintained and reproduced with minimal effort. There are unique foods or parameters described here that make long-term maintenance possible for specific families and orders. However, some types are naturally short-lived or important details needed for them to thrive are still unknown.

Regarding Taxonomy in this Book

Physical specimens were sent for direct identification by a taxonomist whenever possible. Many of the Asian and African species depicted in the text were identified from specimens sent to Dr. Richard Hoffman. A handful of photographs with associated collection data for some North American species were sent to Dr. Rowland Shelley for identification.

Present taxonomic groupings are discussed as relevant to laboratory breeding parameters. The taxonomic information provided here is commonly accepted and likely to be encountered by, and useful to, a diplopod zoologist. This book is specific to living creatures which can prevent taxonomic certainty (i.e. dissection may not be possible and exact geographic origin unknown). Absolute identification of a species typically requires dissection of a mature male for detailed examination of the gonopod structure. In addition, it is usually necessary to have multiple properly-preserved dead specimens of both genders along with precise collection location. This book is dedicated to husbandry parameters and is not designed to tackle taxonomic problems such as difficulties in identifications, synonymy, unavailability of literature, or lack of consensus.

Milliped taxonomy is complex and makes the taxonomic schemes of many animal groups look simple. The seven-tier system of nomenclature has been transmogrified into more than double the tiers when various groupings like subgenera, subterclass, subclass, infraclass, and superorder are included. The complexity primarily reflects the diversity of the class. If the describers' names and dates are included, there exists a six- or seven-word designation without going above genus. One might imagine complexity would lead to improved accuracy, but more misspellings and errors creep in, while synonymy is extremely common. Synonymy occurs when the same species is accidentally described and redescribed with different scientific names. Thousands of species have multiple scientific names. The oldest name holds priority according to the International Code of Zoological Nomenclature and hence is the correct one. The genus can be changed, however, which may require altering the species' suffix to a different gender. *Orthoporus ornatus* has been described at least fourteen times with different names. Various difficulties in milliped taxonomy can confuse even the already confused state of a given species name. The genus *Rhinocricus* is restricted to six species in Cuba and Puerto Rico, but this generic name seems to get tagged onto every Caribbean specimen with a telson spike. *Macrolenostreptus brachycerus* was changed from *Spirostreptus brachycerm* in 1914, but 100 years later millipeds still show up labeled with the wrong generic name (strangely never with the original species name spelling). Temporary taxonomic combinations of two different *Narceus* species (distinct species synonymized) resulted in the reproduction report and parthenogenesis discussion in a later chapter.

(Order Spirobolida) Ring Red Milliped, *Centrobolus anulatus;* KwaZulu-Natal, South Africa
Louis L. Lodder (CC0)

(Order Polydesmida) *Apheloria virginensis;* Montgomery Co., Pennsylvania
Brian Henderson (CC BY 4.0)

(Order Spirostreptida) *Alienostreptus bicoloripes;* Vietnam
Dmitry Ivanov (CC BY 4.0)

(Order Platydesmida) *Gosodesmus claremontus;* San Mateo Co., California
Cricket Raspet (CC BY 4.0)

(Order Spirobolida) *Anadenobolus arboreus* "Red"; Puerto Rico

(Order Polyxenida) *Polyxenus pugetensis*
Multnomah Co., Oregon

(Order Polyxenida) *Polyxenus* face

DIVERSITY

Class Diplopoda is a diverse group of armored terrestrial invertebrates found across the globe. Current estimates are approximately 80,000 species, of which around 10,000 have been named and described. The vast majority perform a vital service to the earth's ecosystems by consuming decaying vegetation and helping its nutrients recycle back into the soil.

According to the fossil record, millipeds are the oldest land animals. One tiny fossil is dated at nearly 500 million years. Even more impressive is fossil evidence that around 300 million years ago some millipeds grew to an estimated six to eight feet in length. Although many people consider the large number of body segments a primitive feature, the most ancient species probably only had around a dozen segments.

Extremely large millipeds can be found in some of the coldest temperate areas, but the greatest variety of species, sizes, forms, and colors occur in the tropics. Millipeds are found in a wide variety of habitats including caves, lava tubes, deserts, plains, forests, grasslands, savannas, and mountaintops. Millipeds can be found inside human dwellings and also in ant nests, where they are kept as pets.

Arthropods include a number of familiar classes that possess jointed appendages and exoskeletons, like insects, arachnids, crustaceans, and centipedes. The arthropod Class Diplopoda comprises the millipeds. According to Shelley (2003) and Shear (2011), this is divided into 2 subclasses, 15 orders, and around 148 families. The first subclass, Pencillata, contains only the Order Polyxenida. Representatives are commonly known as bristly millipeds, have a non-calcified, soft, flexible exoskeleton, lack secondary sex organs, and vaguely resemble tiny, brown dermestid beetle larvae. A truly large specimen only reaches 0.25″ (6 mm). The second subclass, Chilognatha, contains the remaining orders.

- PHYLUM Arthropoda
  - CLASS Diplopoda
    - SUBCLASS Pencillata
      - ORDER Polyxenida
    - SUBCLASS Chilognatha
      - INFRACLASS Helminthomorpha
        - SUBTERCLASS Colobognatha
          - ORDER Platydesmida*—feather millipeds
          - ORDER Polyzoniida
          - ORDER Siphonophorida
        - SUBTERCLASS Eugnatha

Superorder Juliformia
- Order Julida
- Order Spirostreptida*—large round millipeds including *Archispirostreptus*
- Order Spirobolida*—large round millipeds including *Narceus*

Superorder Nematophora
- Order Callipodida
- Order Chordeumatida
- Order Stemmiulida
- Order Siphoniulida

Superorder Merocheta
- Order Polydesmida*—flat millipeds

Infraclass Pentazonia
- Order Glomerida*—pill millipeds
- Order Sphaerotheriida*—giant pill millipeds
- Order Glomeridesmida

Of the various orders, zoologists regularly (purposefully) maintain only the six marked (*), and only the Spirobolida and Spirostreptida with long-term, multigenerational success. At least eight Spirobolida and two Spirostreptida stocks maintained in 2022 date back to 1997-2005.

Of the remaining orders, the diplopod zoologist is certain to run into the *Order Julida*. With around 750 species, this is a medium-sized order, but it is the most ubiquitous. Julidans are almost always brought in with rotten wood or leaves and resemble spirobolidans and spirostreptidans in their rounded, cylindrical (juliform) body form. They are commonly mistaken for babies of the larger orders because many only ever reach 0.5″ (1.2 cm) in length. Julidans at this size may be differentiated because they possess adult proportions. Mature juliforms of all species are thinner and have more body segments than corresponding immatures, independent of size.

(Order Julida) *Ophyiulus pilosus;* Ohio
This is a widespread European species adventive over the eastern United States.

(Order Spirostreptida) *Cambala annulata*
Hocking Co., Ohio

The Julida and Spirostreptida are mostly differentiated by the structure of the lower mouth, while the Spirobolidan can be identified by the variably deep vertical facial groove on the labrum and frons. Mature male julidans usually have hooked or otherwise modified forelegs, but this is hard to see without a stereomicroscope. The chemical defense smells pungent but does not stain skin in the quantity usually produced (although large-bodied, 4″ (10 cm)+ species in the genera *Paeromopus* and *Californiulus* from the west coast of North America produce a visible, yellow exudate). Julidan females commonly form a small cell in wood or earth, deposit a few dozen eggs in a cluster, and stay with them until first instar.

(Order Spirobolida) *Spirobolus walkeri;* China

Order Spirostreptida includes both the world's longest milliped *Archispirostreptus* sp. (15.2″ according to Guinness World Records) and the longest North American millipeds in the genus *Orthoporus*. There are about a thousand species worldwide. Spirostreptidans are predominantly black, brown, or tan, with few exceptions. These are very long millipeds with around 70 body segments. Not all legs are paired and not all segments have legs, so the total is a little less than 280 legs. The fifth segment holds two pairs of legs and the eye patches are triangular or flattened. The defensive exudate contains oxidizing quinones. Both legs on the male's seventh segment are modified into gonopods. A number of species in this order have long captive histories and adults have never been observed molting after maturity—a post-ultimate molt is probably not impossible but would be an oddity. Dozens of spirostreptidans have been bred in captivity and egg laying strategies practiced by this order are discussed in a later chapter.

(Order Spirobolida) *Comanchelus* sp.; Arizona

Order Spirobolida includes the large, colorful, North American *Narceus* species, the gigantic, red-legged, African *Epibolus pulchripes,* and the spectacular Madagascan fire milliped. Spirobolidans are often brightly colored. These also have a highly segmented trunk, but are around twenty segments shorter than spirostreptidans, and so have just under 200 total legs. The fifth segment from the head has only one pair of legs and the eye patches are circular. Like spirostreptidans, defensive exudates contain quinones and both male's legs on the seventh segment are modified into gonopods. However, external structure related to the gonopods is often much easier to see and the male's posterior legs are often oddly shaped and may be used for identification of certain species. Various genera, including *Acladocricus*, *Anadenobolus*, *Apeuthes*,

Aphistogoniulus, Chicobolus, Epibolus, Hiltonius, Narceus, and *Tylobolus,* regularly molt post-maturity (after having obtained full size or produced young), so post-ultimate molts are the rule rather than the exception for this order. These millipeds can be very easy to breed, but female usually have specific egg laying requirements and most eggs cannot survive exposure. Females may lay bare eggs rather than create egg capsules if the correct material is not available (commonly rotten wood fibers).

Order Polydesmida is the largest order with around 3,500 described species. Members include countless spectacularly colorful creatures from habitats in and near the Appalachian Mountains in the United States. Various representatives of the Family Xystodesmidae, particularly the genera *Apheloria, Cherokia, Euryurus,* and *Sigmoria,* are painted in stripes, spots, and squares of glossy red, black, and yellow and are among the mostly brightly colored animals on earth. No matter the color, they fluoresce bright green under ultraviolet light like scorpions. Xystodesmids are only around 2″ (5 cm) in length but are heavy-bodied animals over 1/3″ (~1 cm) wide. The tropics host even larger species like *Nyssodesmus python* from Costa Rica, but they were never imported in the past. Earthy colored *Astrodesmus* and *Rhododesmus,* hardly reaching 2″ (4-5 cm), were sporadically available from Tanzania, Africa, imports from 1990-2005. The Family Paradoxosomatidae contains many small Asian species that resemble tiny, beaded bracelets. The most familiar member, the widespread greenhouse milliped, *Oxidus gracilis,* is sometimes accidentally cultured as a cage pest. In the last few years some related small, colorful species have seen limited husbandry efforts, including *Helicorthomorpha holstii* and *Orthomorpha coarctata,* both common adventives in Florida. Members of this family, even the large, colorful *Antheromorpha,* do not fluoresce under black light.

(Order Polydesmida) *Dicellarius* sp.
Dekalb Co., Alabama

(Order Polydesmida) Mature *Antheromorpha uncinata* (larger) and *Desmoxytes planata* size comparison

Polydesmidan adults often have impressive, flattened dorsal expansions known as paranota extending from each side, plus or minus twenty body segments, and only around sixty legs. While they react to light, they usually have no notable ocelli (eyes) on the head. Many of the large species are brightly colored when mature, but are still pale or colorless as late as subadult. Identifying a species by color before the ultimate molt is often impossible. Only the front set of the male's leg pairs on segment seven is modified into gonopods. The defensive secretion smells deceptively pleasant, like cherry or almonds, because it contains benzaldehyde along with poisonous hydrogen cyanide. Members of this order are sometimes confused with centipedes because they have square, flat body segments, but polydesmidans have two pairs of legs per segment and do not have maxillipeds (venomous claws surrounding the head of centipedes). Adults are often short-lived, a few months to a few years, and post-ultimate molts do not occur.

Order Sphaerotheriida rarely exists outside the Old World tropics. They are large, stout, ovate animals able to roll into a perfect ball or sphere when disturbed. The body is composed of thirteen segments (12 visible when curled up) and the leg count is commonly 40-44. Males do not have gonopods, but the last one or two pairs of legs are modified into clasping, claw-like structures called *telopods*. Sphaerotheriidans do not secrete a defensive fluid when handled but stay curled up a long time. The most famous representatives are the giant green pill millipeds from Madagascar, once popular as both pets and swarms in nature documentaries. Probably multiple species and genera were regular, seasonal imports by the late 1980s, some as long as 3.5″ (8.9 cm). An Indian species up to 1.5″ (3.8 cm) from Tanzania became a common import starting in the late 1990s and some similar-sized species have been imported from the Philippines and Thailand. Limited husbandry success suggests post-ultimate molts may be normal.

Order Glomerida includes hundreds of warningly colored species, mostly from Europe, northern Africa, and Asia, that were unknown in captivity (outside of endemic research locally sourced) prior to 2015. *Glomeris pustulata* and *G. pulchra* have seen multi-generational success from 2017. Some offspring have been produced by other *Glomeris spp.* and *Rhopalomeris carnifex*, but without evidence of a subsequent generation. Members of this order vary in size; those in the United States and Mexico are tiny, similar in size to the isopod crustaceans known as pill bugs. *Onomeris* species found in the southeastern U.S. are difficult to acquire and collect, possibly due more to them being difficult to see, than to limited distribution. Glomeridans have defensive glands that produce a sedative similar in structure to quaaludes. As a sphere they appear to have only eleven segments since segments two and three are fused. Males have claw-like structures on the terminal legs for mating. Mature specimens often continue to molt, about twice a year.

Diverse arthropods volvate into spheres as a means of defense, not just members of the three pentazonia milliped orders. Humans often confuse pill millipeds with the commonly encountered pillbugs (terrestrial isopods). They have the same physical defense and accordingly a superficial resemblance. However, a close look exposes countless differences including the crustacean 'tail,' the multi-segmented abdomen, and uropods. Isopods are crustaceans like lobsters and shrimp. Isopod mothers carry eggs and early developmental stages under

(Order Glomerida) *Glomeris klugii;* Italy

the body within a specialized pouch formed by oostegites. Pill millipeds bury the brood—they cannot carry eggs or larvae. Other creatures with a similar defense are found among the insects in the Order Blattodea. The pillbug cockroaches in the genus *Trichoblatta* are among the most perfect sphere formers. They look similar to pill millipedes from above, but there are only six legs beneath. Pillbug cockroaches do not lay eggs: females retract oothecae into the abdomen until the young are ready to emerge. Although the pill-forming invertebrate groups mentioned here look identical to the average person, the way the segments

Pillbug cockroach, Japan

(Order Sphaerotheriida)
Arthrosphaera brandtii
Tanzania

(Order Platydesmida)
Brachycybe producta
California

(Order Polyzoniida)
Petaserpes sp.
Hocking Co., Ohio

are shaped and the way they come together in the defensive curl is quite different. There are other examples, including a few species of flat millipeds (Polydesmida) built to roll into balls, certain beetles, and now-extinct trilobites, as well as lizards and mammals that retract into semi-spheres.

Order Platydesmida contains small but beautifully structured species which sport spectacularly elongate paranota. The paratona contain mostly defensive glands. When disturbed, specimens exude defensive fluid containing a novel alkaloid. The exudate is either very rarely used or barely detectable to the average human (though it may greatly deter ants). The pinnate appearance created by the long paranota has earned this order the common name feather millipeds. Only about sixty species are known worldwide. A few native North American species are seasonally available from specialty vendors, especially *Brachycybe lecontii. Brachycybe* species produce young in captivity and are maintained over short periods with some success. Males guard the eggs and small young. They are gregarious and found in groups that persist over years and contain varying stages of development.

Order Polyzoniida includes seventy or so species worldwide and is in the Subterclass Colobognatha along with the Order Platydesmida. These are small and elongate with a domed body and flat underside. Common types in North America include *Petaserpes* species which are pale pink, skinny, and around 0.75″ (~2 cm) in length. They are sensitive to warm temperatures and difficult to keep alive for more than six months.

(Order Callipodida) *Abacion magnum*
Dekalb Co., Alabama

Order Callipodida includes velvety *Abacion* spp. which can exceed two inches (5 cm) in length. They are long and thin and superficially resemble Juliformia. The hundred and thirty or so members of this order excrete stinky phenols in defense. Adults are on the large side of tiny and look interesting, but they are seldom available commercially, so a researcher interested in studying their captive husbandry may need to collect them.

Order Chordeumatida contains species that rarely exceed 1″ (< 2.54 cm) but can move like lightening, as fast as a house centipede or cockroach. These have 26 to 32 body segments. They are somewhat thick-bodied and commonly known as sausage millipeds. While these make up the second largest order with around 1200 species, I have only seen a single specimen.

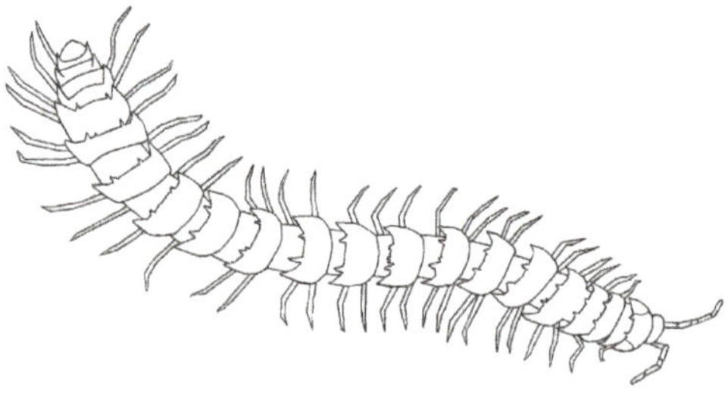

(Order Callipodida) *Abacion* sp.; Canton, Georgia
Christina Butler (CC BY 2.0)

(Order Callipodida) *Abacion* sp.; Washington, D.C.
Katja Schulz (CC BY 2.0)

(Order Polyzoniida)
Buzonium crassipes
Humboldt Co., California
Casey H. Richart (CC BY 4.0)

(Order Polyzoniida)
Petaserpes cryptocephalus
Richland Co., Ohio
Brad Von Blon (CC BY 4.0)

(Order Polyzoniida)
Rhinotus purpueus
Lee Co., Florida (introduced)
Paul Marek (CC BY 4.0)

(Order Polyzoniida)
Octoglena bivirgatum
San Mateo Co., California
Cricket Raspet (CC BY 4.0)

(Order Polyzoniida)
Siphonotidae
New Zealand
Aimee Macarthur (CC BY 4.0)

(Order Polyzoniida)
Stenozonium exile next to pine needle
Humboldt Co., California
Paul Marek (CC BY 4.0)

(Order Chordeumatida)
False Flat-Backed Milliped
Nanogona polydesmoides
Ukraine
Igor Balashov (CC BY 4.0)

(Order Chordeumatida)
Striariidae
Clackamus Co., Oregon
Jeffrey Lee (CC BY 4.0)

(Order Chordeumatida)
Tingupidae
British Columbia
Rambryum (CC0)

(Order Chordeumatida)
Connecticut
Christine Young (CC BY 4.0)

(Order Chordeumatida)
Craspedosoma rawlinsii
Austria
Georg Haberfellner (CC BY 4.0)

(Order Chordeumatida)
Schedotrigona sp.
New Zealand
Tyler McBeth (CC0)

(Order Polyxenida) Canberra, ACT, Australia
Donald Hobern (CC BY 2.0)

(Order Polyxenida) San Juan, Bolivia
Kozue Kawakami (CC BY 4.0)

(Order Polyxenida)
Polyxenus lagurus
Berlin, Germany
Alexis Orion (CC BY 4.0)

(Order Polyxenida)
Idaho
Jake N. (CC BY 4.0)

(Order Polyxenida)
Polyxenus lagurus
Hungary
Sergey Khm (CC BY 4.0)

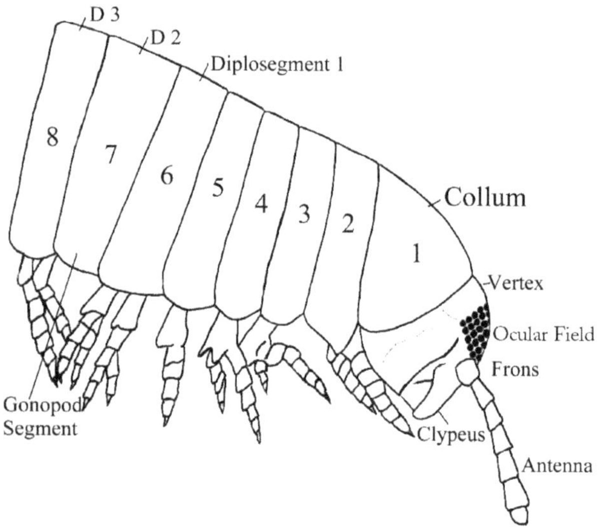

Fore body, *Narceus americanus*
Some researchers believe the front segments are reduced diplosegments. (McMonigle)

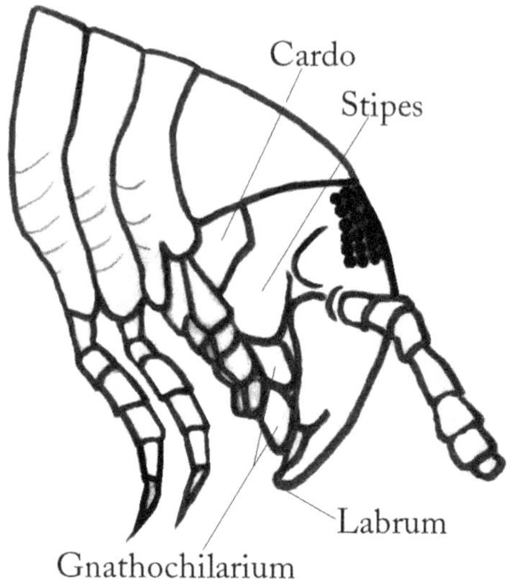

Mouthparts, after *Floridobolus orini*
Marion Co., Florida (McMonigle)

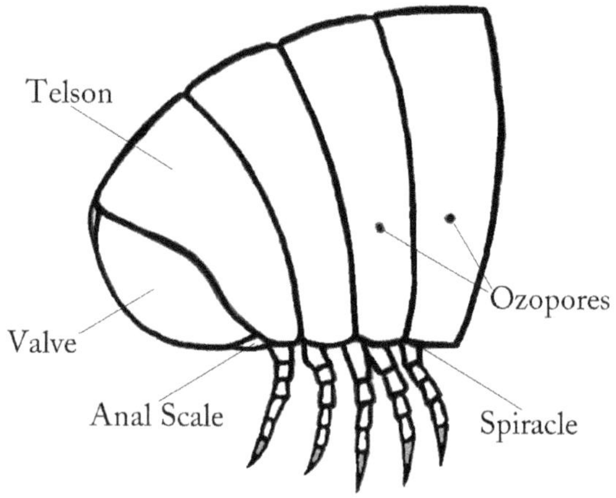

Telson and valves, after *Floridobolus orini*
Marion Co., Florida (McMonigle)

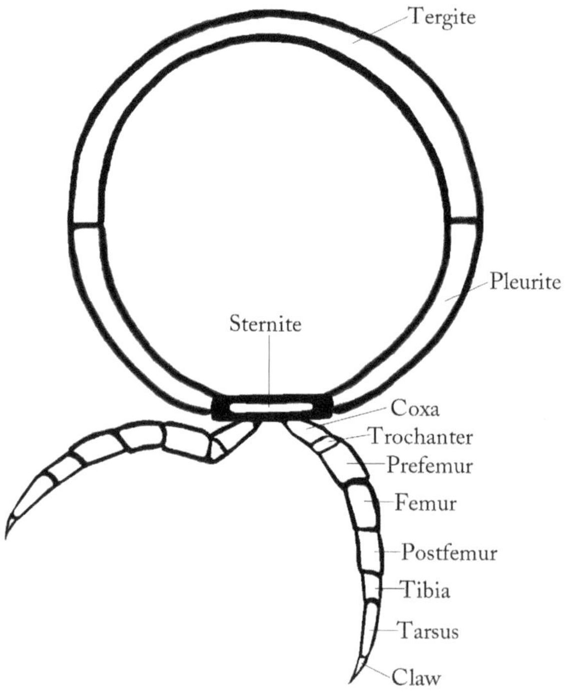

Generic Juliformia leg and body segmentation (McMonigle)

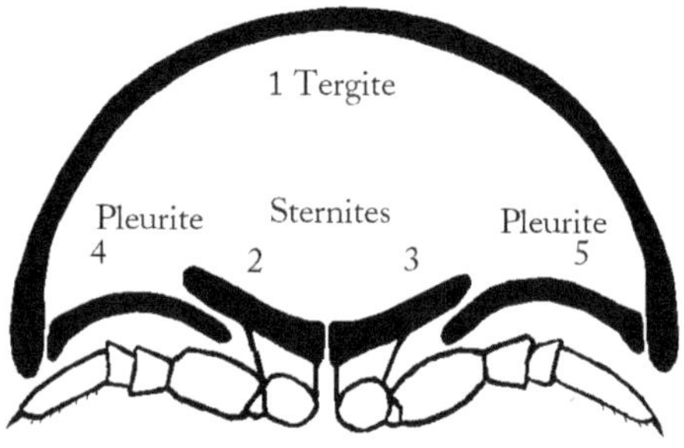

Each Pentazonia body segment is composed of five zones, after Arthrosphaera. (McMonigle)

MORPHOLOGY AND BIOLOGY

The most familiar characteristic of Class Diplopoda is the large number of legs. 'Miliped' translates as a thousand feet, an explanatory exaggeration referring to the seemingly countless rows of legs. Most have forty to three hundred legs and a true thousand-footed animal was imagined only as hyperbole. For almost a century, one tiny species, *Illacme plenipes* from California, was thought to have the most legs, up to 750 (Cook and Loomis 1928). A more recent discovery from Western Australia, *Eumillipes persephone* Marek, 2021, has as many as 1,306 legs.

Animals in the Class Diplopoda have a head and trunk. The head contains two antennae, usually chewing mouthparts (some are modified into sucking mouthparts) and usually two 'eyes', each made of a variably shaped cluster of ocelli. The trunk usually has four haplosegments and a dozen to many hundreds of 'apparent segments' or diplosegments. Each apparent segment has four legs or two pairs of legs (diplo = 2, poda = foot). The first segment behind the head (the collum) is often large and without legs. Most body segments are more accurately called diplosegments because, though each looks like a single segment, it is formed from two fused segments.

As a group, millipeds have chewing mouthparts and are considered detritivores. There are peculiar exceptions including species that siphon juices from living plants and others that attack and eat small animals. However, these exceptions are tiny oddities with little to no husbandry parameter or life cycle documentation.

The face has the vertex at the top, frons in the middle, and clypeus near the bottom. The labrum is the upper part of the mouth, the gnathochilarium is the bottom, while the cardo and stipes form the basal sides of the mandibles which terminate in gnathal lobes. The first body segment, known as the collum, may be enlarged or reduced in size. The second segment contains the penes in the male or ova ducts and seminal receptacles known as vulvae in the female. Vulvae terminate in sacs called spermathecae which store the spermatophore. In males of most orders the seventh segment contains secondary sexual organs developed from the legs called gonopods. The rear legs form the spermatophore from the penes on the second segment and place it in the gonopods. The terminal segment is known as the telson—beneath is a pair of anal valves (paraprocts) with a small anal scale (hypoproct) located at the bottom.

Sensory apparatus is primarily geared for direct contact as the antennae and legs contact soil while the body responds to touch. There can be a number of ocelli in the ocular field—so much that they vaguely resemble the compound eyes of insects, but there is usually a thick wall between each ocelli. The Pencillata specifically can have defined compound eyes. Some millipeds lack eyes altogether. Species that appear to lack visible ocelli are often capable of distinguishing light from dark through reduced ocelli or light detecting organs on the antennae. There is also a sensory pit between the antenna and ocular field called the Tömösvary organ which may aid the ability to detect air pressure, sound, odor, and humidity.

The milliped respiratory system is similar to that of insects and centipedes. Air enters through the spiracles and diffuses along a network of branching tracheae connecting to each body cell. Unlike many centipedes and insects, the spiracles are not located along the sides of the body. They are very close to the base of the legs and are usually visible only with a hand lens.

Despite the similarity to insect and centipede respiratory systems, millipeds possess a unique ability: they do not easily drown. Even many aquatic insects drown if held under water. Various Amazonian millipeds are known to survive submersion during flood periods, but this is not a restricted ability. About twenty years ago I woke to a very depressing sight. A chiton blocked up the return pipe to my reef aquarium and its body caused the water to reflect under the stand and overflow a 10-gallon terrarium containing about sixty *Narceus annularis*. I was rather depressed about all my dead millipeds—these were my babies, and most were finally approaching maturity after I had been caring for them more than three years! Sad and with no desire to sort through dead millipeds, I delayed dumping the cage. I was feeding some other inverts the next day and out of the corner of my eye I could see millipeds performing slow-motion somersaults in the water. I thought I was seeing things but eventually took a closer look. They were 'running' underwater! They had been submerged in saltwater 24 hours. I ran the tank outside and dumped it in the driveway. Only two were dead. I rinsed the survivors and the cage with fresh water, replaced the substrate, and they lived for years.

Pigmentation ranges from shocking warning colors to basic camouflage or pale translucence. A number of different pigments are involved in coloring the exoskeleton.

Body ring shape:
Floridobolus (two on left)
Chicobolus (two on right)

Pigments occurring in diplopods include flavins (yellow and red), melanins (brown and black), tetrapyrroles (yellow and green), and quinones (blue and purple). Pigments and coloration are not always straightforward—there are additional pigments found primarily in arthropod eyes known as ommochromes. Most diplopods lack pigmentation when they are very young, commonly followed by intermediate juvenile and distinct adult colorations. Three distinct color phases relative to development are common, though many xystodesmids are pale to the last molt. Cave-dwelling species usually lack all pigments, while the rare albino is not a pale form, but a variant of a normally dark-colored species. Albinos are unable to synthesize melanin but are still capable of synthesizing other pigments. Due to chemical changes, bright pigments usually change greatly when dried, so even the most beautiful milliped may become brown and drab as a specimen.

Exoskeletal features of many species fluoresce under ultraviolet light. The most spectacular are the large flat millipeds, but a number of round species also exhibit this phenomenon. Every part of the body and legs of various members of our large native polydesmidans, including *Apheloria, Euryurus, Harpaphe*, and *Pachydesmus,* glow a vivid neon green like scorpions. A few small spirobolidans such as *Spirobollelus* and *Benoitolus* also glow bright green over the whole body. The body segments of some large round millipeds like *Archispirostreptus gigas, Dendrostreptus macracanthus, Tylobolus uncigerus,* and *Trigoniulus macropygus* glow a subtle forest green. The legs of *T. macropygus* also fluoresce, but neon orange rather than green. Orange is an unusual fluorescence in living arthropods, but the antennae of *Anadenobolus arboreus* 'red' also fluoresce orange. The body of *Acladocricus* and the legs of *Anadenobolus arboreus* from the Virgin Islands glow blue under 350 nm. At a glance it seems that most captive species do not fluoresce, but on close inspection the tiny tips of the claws of nearly every spirobolidan and spirostreptidan fluoresce.

Bioluminescence is extremely rare, but has been identified in two very different orders. A tiny species from the Order Spirobolida, *Paraspirobolus lucifugus* (Gervais, 1836), is bioluminescent, glowing along the front edge of each segment. This small species has been encountered in greenhouses across Europe. It was introduced by humans to most tropical areas across the globe including Africa, Australia, tropical Americas, Asia, and India, so its natural origin is uncertain. Far more massive creatures from the Order Polydesmida, at least nine species in the xystodesmid genus *Motyxia* from southern California are bioluminescent. The entire body and legs glow, which is unusual since other groups like worms, beetles, centipedes, and brittle starfish display bioluminescence in only a portion of the body. These partial body displays are often very bright but fleeting. On the other hand, milliped bioluminescence is continuous but dim, only visible in extreme darkness.

Like all arthropods, millipeds must shed the exoskeleton as they grow. Weeks prior to the molt they construct a protective molting cell in the substrate. Most molting cells are simply scraped out caverns beneath wood and rocks, or deep in the substrate, and are only as stable as the surrounding substrate. Polydesmidans (flat millipeds) build an amazing little frass igloo that will shelter them during the impending molt. Glomeridans

(Order Polydesmida)
Pachydesmus crassicutis
Louisiana
Fluorescence under UV light

(Order Spirobolida)
Spirobollelus immigrans
Hawaii
Coloration under daylight versus ultraviolet

(Order Polydesmida) *Apheloria virginiensis;* Ohio
Subadult coloration under daylight versus ultraviolet

(small pill millipeds) also construct solid-walled molting cells. All species are extremely vulnerable during and after the molt. It is incredibly easy to dent and tear the new exoskeleton. Freshly molted specimens can die from excessive handling, even if visibly undamaged. If the cell collapses during a vulnerable period, it will crush the milliped. It takes a few days to a week for the new shell to harden, with older and larger specimens and thicker exoskeletons taking the longest. It can take many weeks for maximum rigidity, so care should still be taken when handling a milliped that has molted in the last month. Molts can be difficult to track, but approximately one to two dozen are needed for various species to reach maturity.

The molting process from start to finish takes from a few weeks to more than four months. In deep, moist substrate, millipeds dig down and construct a molting cell. If the substrate is too loose or shallow in the artificial habitat, and the animal cannot form a cell, it will eventually begin the internal, transformative process. Unless there is something harmful in the habitat, they molt just as well on the surface since they are not subject to predators or rough weather. All species become mostly immobile as a new exoskeleton forms beneath the old one. The transforming specimen remains in a loose curl (while a limp and unnatural position is a sign of death rather than molting). If a milliped appears white or chalky, it should be placed on the surface. It should never be reburied. It may not have the strength or time to construct a new cell. When ready, the old shell splits open and is cast off within a few minutes to a few hours. The new carapace is puffed up with fluids and then can begin the slow process of hardening.

Keepers must be very careful when digging through deep substrate because freshly molted millipeds are easily damaged or torn in half. In some species, especially spirostreptidans like *A. gigas,* the molting cells are often clustered in one area so if one is disturbed it is often part of the wall of the cell next to it. When molting cells are accidentally disturbed during substrate maintenance and replacement, it is usually safest to isolate the specimen. Uncovered specimens should be placed on a 0.25″ (6 mm) bed of damp substrate in a container with minimal ventilation. Millipeds molt successfully on a flat surface—a shallow substrate retains moisture while preventing incomplete attempts to rebuild the cell. If the disturbance is early, the animal will attempt to rebuild a molting chamber before settling down again to molt. Increased sluggishness and discoloration of

This *Euryurus leachii* (Order Polydesmida) is constructing a molting cell with frass droplets. Every molt is preceded by a considerable building effort.

(Order Sphaerotheriida)
Arthrosphaera brandtii
Freshly molted animal and exuvium

(Order Spirobolida)
Acladocricus sp.
Negros Island, Philippines
The old exoskeleton becomes highly discolored days prior to the molt.

(Order Spirobolida)
Acladocricus sp.
This is a freshly molted animal and discarded exuvium. Note, the split occurs between the sternites and pleurites.

the exoskeleton indicates the animal probably will not be able to form a new cell.

Millipeds take on a gray or milky appearance a few days to a few weeks before the molt. Premolt specimens tend to stay in a tight coil—if a discolored animal is limp and not in a coil it is probably dying. A fluid that separates the new and old exoskeleton causes discoloration as it begins to break down the old exoskeleton. The color change is obvious in most species. In round and flat millipeds, the shell splits along the sides of the segments between the tergites and pleurites or near the base of the legs, from head to telson. The cast-off exoskeleton is off-white, thin, and frail. The exuvium is abandoned following surface molts (if the molt is observed the exuvium will not be eaten) but seems to be eaten following concealed molts, since the lower substrate layer is not littered with exuvium. Pill millipeds exit through a split behind the head and leave behind a nearly intact and fully colored exoskeleton.

Milliped segments are hardened, highly sclerotized bands—with exception of the Pencillata. There are no elastic side (pleural)

(Order Glomerida)
Glomeris pulchra molting cell uncovered

regions like those found on centipede trunk segments or arachnid and insect abdominal segments, just hardened pleurites. For this reason, growth between molts occurs at the segment joints rather than between the top and bottom of a segment, somewhat like the expansion of a telescope. This is the reason a freshly molted milliped is shorter, sometimes half as long as it will be following a few weeks of feeding. This also makes measurement of adults a somewhat fluid proposition. A large, wild-caught *Orthoporus* can expand nearly an inch (2 cm) after twenty minutes of feeding on melon. *Thyropygus allevatus* can elongate twice that amount. In the Pentazonia, so named because the five parts of the band are not fused (1 tergite, 2 pleurites, and a 2-part sternite), expansion and contraction occurs along the bottom four plates. For these pill millipeds, the weight of the animals or frass production are the clues used to determine food intake as healthy animals roll up into a sphere if an attempt is made to check the underside.

The protonymph is a peculiar developmental stage that occurs in the first stadia of Juliformia, as well as centipedes, and is akin to the post-embryo stage of arachnids. The egg hatches into a lightly segmented, croissant-shaped tube, with rudimentary appendage buds. The protonymph cannot feed and cannot go anywhere (it is usually motionless but can move its head and flex when it approaches the end of its development). This is not counted as an instar because it is a unique stage (just as eggs, insect pupae, and sexually mature specimens are normally not counted among larval instars). The protonymph splits the eggshell and can take days or weeks to fully emerge from the shell. It enlarges as it develops and after two to four weeks the eyes of the first

(Order Julida)
Californiulus yosemitensis
Molting cell hollowed out under a clump of moss (moss removed)

(Order Spirostreptida)
Telodeinopus aoutii
One on top is molting and one on bottom will soon molt. Note the split takes place between the tergites and pleurites and the premolt is not obvious as in *Acladocricus*.

Molts completed. As is usual for exposed molts, the exuviums were not eaten.

(Order Spirostreptida)
Earthworm Milliped
Macrolenostreptus brachycerus
(Top and bottom) Molting cell scraped out under a piece of wood.

instar darken beneath the protonymph's skin. Defined body segmentation also begins to show through. Development is easy to view because the protonymph's surrounding membrane is thin and transparent. It is interesting to note the protonymph doubles in size after hatching and the first instar immature can be half as much larger again, all without feeding.

The first instar larva emerges with a functional mouth and legs. However, it is still pale white and is different from later instars because only the first few segments have legs. (In Juliformia there are normally six legs at the front of each side and an ozopore near the middle—see top photo p. 49.) This stage burrows, feeds, and then molts to second instar. Second instar is a pale version that now looks like the adult but there are fewer body segments and ocelli (when present). Additional body rings and legs are added with each molt until the adult number is reached after seven to ten stadia.

Spirobolidans and spirostreptidans vary by species and husbandry in the time required to reach maturity. Under optimal conditions the majority need at least three years to grow from egg to adult. The quickest cultured species is *Anadenobolus monilicornis* which can be sexually mature in twelve months, though it still takes three or four years to reach maximum size. *Trigoniulus macropygus* takes sixteen to twenty-four months to mature, but it is a far larger creature and does not reach sexual maturity before attaining maximum size. *Archispirostreptus, Spiropoeus,* and *Pelmatojulus excisus* are among the slowest and can spend as long as seven years to reach maturity. Size is not the determining factor as small *Spirostreptus* species can take a few years longer than the gigantic *Epibolus pulchripes.*

(Order Sphaerotheriida)
Madagascan *Zoosphaerium* beginning to molt (top) and following completion (bottom)

Giant round-millipeds are normally long-lived. The big spirostreptidans and spirobolidans take three to five years to reach adulthood and then can live another fourteen years after reaching sexual maturity (McMonigle 2012b, 2018b). The total lifespan of many members of these orders can exceed a decade, but some only live a few years. African *Centrobolus* and North American *Chicobolus* spirobolidans commonly live only three years, while the Puerto Rican *Anadenobolus* seem to be naturally short-lived, seldom more than a year after reaching adulthood. It is possible they are longer lived in the wild than in the laboratory. Longevity is not always consistent for the same species, stock, or generation. In 2005, half a dozen wild adult Negros Island *Acladocricus* died in less than a year, and specimens of the F1 generation expired within eighteen months of reaching adulthood. Specimen longevity increased intermittently over subsequent generations and some specimens of the fourth generation lived as long as six years after reaching maturity. Although giant pills (spaerotheriidans) live a few weeks to a few years in artificial habitats, the lifecycle may span a decade in nature. Small pills (glomeridans) take two to three years to mature, but then only live a year or two. The large polydesmidans (flat millipeds) rarely live a year in captivity, while the life cycle may take two to four years in the wild. Many are seasonal breeders like the Euryurini (and probably all paradoxosomatids) and do not normally exceed two years from hatching to old age.

Wild-caught specimens can have their lives cut short due to importation stresses or damage from collection methods. Sphaerotheriidan pills and heat-intolerant flat millipeds are likely to die within a few days to a few weeks of arrival. Old and molting specimens are especially sensitive to rough handling, temperature extremes, long periods in shipping, and poor ventilation. *Archispirostreptus* tend to ship successfully, but one imported specimen was damaged in such a way that its mouthparts would not move. It could not eat but lived for four months. Millipeds not only live slowly but also die slowly. A limp or inactive milliped is as good as dead even though it might survive for months. Pill millipeds that do not rapidly roll into a tight ball will smell of rotting milliped in short order. Dying animals do not always show signs of difficult shipping. If there are a few dead animals in a group of imports the rest can sometimes all be dead in a few weeks, even though they appear active and healthy.

Nearly all groups of large arthropods have their own species of commensal mites—millipeds are no exception. Imported *Archispirostreptus gigas* almost always have a few dozen, fast-moving mites living on their body. These mites are many times larger and faster than the commensal mites encountered on the familiar hissing cockroaches. *Archispirostreptus* mites are impressive because, despite their speed, they seldom leave the milliped. They move so quickly it seems they should run right off the host. The African *Dendrostreptus macracanthus* have very different commensal mites. Theirs are small, slow moving, and hang out on the lower sides of the segments, just above the legs. The Philippine *Acladocricus* hosts two large species: a lumpy, orange species with normal legs that prefers to cling to the milliped legs and another with antenniform front legs that lives on the body. Another big southeast Asian species, *Thyropygus allevatus,* sometimes hosts large, fast moving mites with long front legs.

(Order Spirobolida) *Trigoniulus macropygus*.
Legs move in waves. Speed can be estimated by the number of waves.

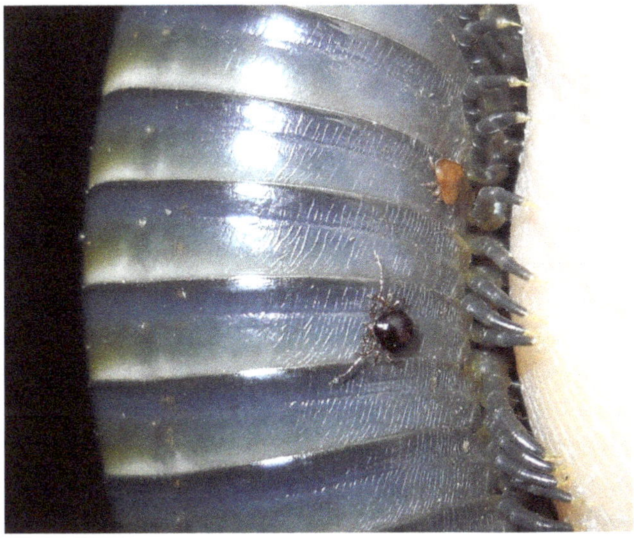

Narceus gordanus are uncommonly found with large, dark mites that move slowly and orient themselves above the legs. The majority of milliped species do not have a commensal mite species, or as with *Narceus annularis* (host to two endemic mite species), are almost never encountered with them.

Commensal mites do not cause harm to their host, and help keep the body clean of mold and hypopus mites, but some keepers prefer to get rid of them. Removal consists of carefully pulling the mites off by hand. It usually takes three or so inspections before the mites are eradicated. Some hobbyist feel the mites are beneficial and leave them be. However, it is difficult to keep commensal mites alive in captivity long term, which is why captive-bred specimens almost never host commensals (unlike hissing cockroach mites which reproduce spectacularly in captivity as long as there are host roaches).

Mass migrations of millipeds are most commonly reported for polydesmidans, but occur in other orders. The aggregations are often seasonal and usually consist of adults but there are many variations. In Japan, massive swarms of 'train millipeds,' the polydesmidan *Parafontaria laminata*, can build up on railroad tracks and stop trains in their tracks as the smashed bodies prevent the wheels from gripping the rails (Hopkins

(Top) Commensal mites on *Thyropygus allevatus*
Thailand

(Middle) Commensal mite on *Thyropygus* sp.
Thailand

(Bottom) Two types of commensal mites found on *Acladocricus* sp. (Negros Island, Philippines) are shown. The dark species with long front legs is a member of the family Heterozerconidae, genus *Allozercon*.

Narceolaelaps gordanus (commensal mites) on *Narceus gordanus*
Marion Co., Florida

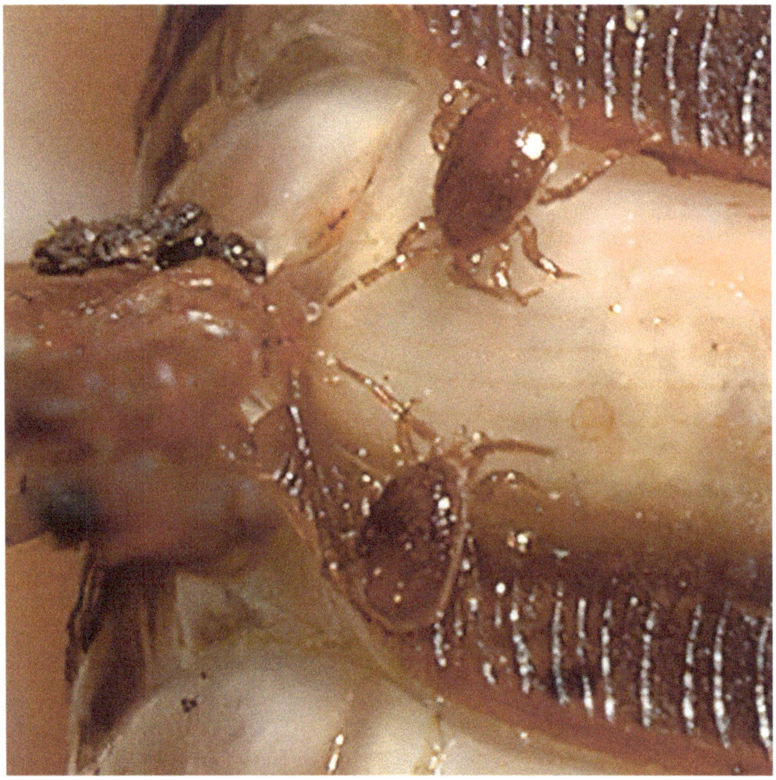

Julolaelaps sp. (commensal mites) on *Archispirostreptus gigas*

Phoretic mites on scarab larva look the same on millipeds, but I have avoided the problem for some decades and have no pictures. This is a harder-to-avoid problem on scarab larvae. For discussion of pest mites and millipeds, see pp. 65-67.

The three red polyps near the front are parasitic *Hydryphantes* sp. mite deutonymphs on a male *Belostoma* sp. water bug. (The white and gray objects are eggs laid by the female water bug.) The attached deutonymph on the upper left is molting into an adult water mite. The adults mites are excellent swimmers.

Leptus sp. mites that parasitize some polydesmid millipeds look the same as those on this opilione.

and Read 1992). In the southeastern United States, sporadic aggregations of many dozens of southwestern *Orthoporus* are reported (J. Hiykel, pers. comm. 2011)—since they are adults, a congregation pheromone for mating is likely involved. In the northeastern United States, large migrations of wild *Narceus* are reported from time to time. Groups of hundreds and hundreds of adults are observed fleeing from an area of the forest (J.T. Bowman, pers. comm. 2005). Where they are going and what stimulus prompts this uncommon behavior is a mystery. They may run out of food in an area, but how does a large group of normally solitary creatures coordinate unidirectional movement?

Diplopods are commonly gregarious, and some, such as *Narceus annularis,* occur on large rotten, fallen trees with every stage of development (eggs, immatures from one- to two-years-old, and adults) found concurrently in midsummer (pers. observ.). However, members of this class are considered solitary animals with exception of some members of the Order Platydesmida, the feather millipedes (Wong 2017). *Brachycybe* specimens of various stages and genders cluster and feed together in groups year after year, with adult males guarding the eggs. Most species in other orders abandon the eggs, but a mother spirostreptidan *Dendrostreptus macracanthus* guards eggs through the first two molts, while *Pseudopolydesmus* and some julidans wrap their bodies around the eggs.

Defenses Physical and Chemical

Millipeds are harmless to humans and do not bite, pinch, sting, or carry disease. What they lack in offense they make up in defense. Diplopods are foremost among arthropods in the development of defensive body armor and arsenals of chemical deterrents. They are the ultimate defenders, built to ward off all but the most determined predators.

The thick, hardened exoskeleton is used as the first line of defense. The defensive coil is a behavior found among most members of the various orders that allows them to use the armor effectively. The head and legs are brought inward and protected by the hardened upper-body plates. All the pentazonians and some polydesmidans take the coil defense to its ultimate culmination by transforming into an impenetrable sphere. Although possession of a coiling defense is nearly universal to the diplopods, the duration and associated behavior can vary greatly. Some millipeds will remain in the defensive coil for half an hour, while others grow impatient within seconds. Most slowly uncoil or unroll—others like *Floridobolus orini* 'Ocala' uncoil rapidly like a spring. Some species partially coil and uncoil rapidly when sufficiently annoyed. (*Dendrostreptus macracanthus* and some *Narceus gordanus* do this.) The rapid flopping is often accompanied by toxifabricant release. The defensive coil is not always about predators, some julidans coil during flooding events and float away to safety.

One of the class features of the Diplopoda is the presence (in most species) of glands, opening through laterally placed pores, able to secrete a variety of volatile, toxic organic compounds. Diplopods manufacture toxic compounds from simple

(Order Spirobolida) *Tylobolus uncigerus*
Defensive fluid from ozopores on rear half of body

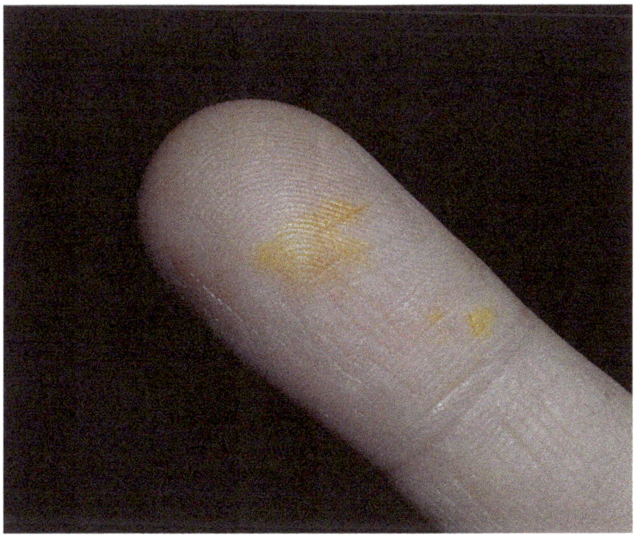

Initial contact with an exudate

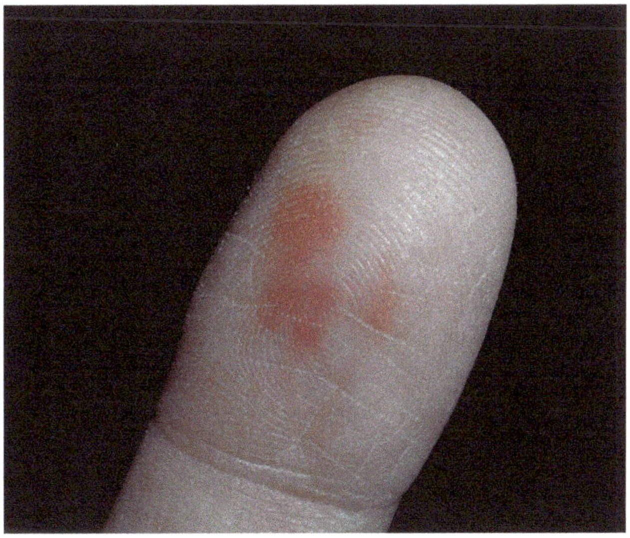

After twenty minutes

ingredients obtained through their normal diet of leaf litter, starting with simple sugars and amino acids derived from the digestion of plant tissue. Glands manufacture chemical defenses—ozadenes are the glands (Greek, *ozos,* "a bad smell," + *adenos,* "gland") while ozopores are the discharge sites. The scientific names of a number of species reference the chemical secretions: *Abacion lactarium* (Say, 1821) references the milky color of the secretion and *Callipus foetidissimus* (Savi, 1826), alludes to the disagreeable odor of that species (said to smell like decomposing sponges). While most simply ooze defensive fluids, some can eject it long distances. *Rhinocricus latespargor* ("far spitter") from Haiti discharges jets of defensive secretion up to thirty-three inches on each side of its body and can do so repeatedly. *Aulacobolus ejaculans,* discovered in southern India, also ejects secretions under pressure (Hoffman 2008).

Even such lethal mixes as phenols and hydroquinones are manufactured and stored in spherical, thin-walled ozadenes, discharging to the exterior through a short duct having a muscle-operated closure valve. Some spirostreptidans and spirobolidans are capable of producing a quantity of strong defensive fluids but are immune to high concentrations. Polydesmidans have a two-part system to avoid self-intoxication with hydrogen cyanide. Mandelonitrile is produced in a chemically inert precursor form in an inner chamber lined with secretory cells. For use, it is extruded through a valve into an outer reaction chamber with impervious lining, where it is catalyzed by an (unidentified) enzyme to produce the creamy exudate containing hydrogen cyanide and benzaldehyde that gives xystodesmids their pleasant odor. The pleasant odor betrays its lethal ability: small mammals confined with such a milliped may be quickly dispatched. However, polydesmidans disturbed inside enclosures with very limited ventilation rarely, if ever, succumb to their own poisons.

Some orders have no chemical defenses. Defensive chemicals are absent from sphaeriotheriids, polyxenids, and chordeumatids (never present in these groups), and secondarily lost in many polydesmoids. In these, and the well-known 'juliform' species (spirobolidans, spirostreptidans, julidans), the glands may occur on all segments from the fifth to the penultimate, one on each side. In glomeridans, the pore is located middorsally, and produces quite a different kind of secretion. Glomerin has been found to induce sedation in predatory spiders and the fluid is sticky like syrup. *Petaserpes* produce polyzonimine, which is known to be a superb defense against army ants. Many of the platyxenids produce deoxybuzonamine,

(Order Spirobolida)
Anadenobolus monilicornis

which is an alkaloid related to quinine, caffeine, and morphine.

Milliped chemical defenses should be taken into account during handling of specimens. Wild specimens commonly secrete protective fluids with little provocation. Animals that have recently molted secrete readily and can produce copious amounts of fluid within the first few days. Under normal circumstances, many species only secrete the fluid when handled roughly, but some like the African *Dendrostreptus macracanthus* can be difficult to handle gently enough to prevent secretion. Cultured specimens and species that refuse to release defensive fluid during regular handling will still react if sufficiently molested or damaged.

Although the active chemicals can include various quinones, the defensive fluids of spirostreptidans and spirobolidans have the same properties. The exact concoction will vary by species, age, and individual specimen. The chemical smell is pungent and the fumes cause one's eyes to water in close proximity. The fluid stings if it gets into the eyes or an open cut. It should be washed off quickly. A few minutes after the yellow fluid contacts human skin it can cause reddish brown stains that do not wash off. The color of the affected skin turns red-brown, which only fades away after a few days.

Despite size and defensive secretions, predators eat the largest and most protected millipeds. Gigantic *Scolopendra* centipedes are common in most of the same areas and make millipeds a portion of their natural diet. Giant vinegaroons also count millipeds among their prey. An entire subfamily of assassin bugs (Reduviidae; Ectrichodiinae) with more than 600 species, feed only on millipeds, and mostly just the big spirostreptidans and spirobolidans. At least one dung beetle species from South America, *Deltochilum valgum*, feeds on dead or dying millipeds. The beetles from the Family Phengodidae (the milliped hunters) are also very specialized milliped predators that eat nothing else. Adult females and larvae are difficult to tell apart because they are long, thin, and wingless. North American milliped hunters feed mostly on large *Narceus* and *Orthoporus*. They do not eat small millipeds or any other invertebrates.

(Order Polydesmida)
Antheromorpha rosea

(Order Glomerida) *Glomeris klugii*
Glomerin exudate as clear droplets on the right side of the dorsal surface

(Order Polydesmida)
Pachydesmus crassicutis
Lousiana

(Order Glomerida)
Hyleoglomeris sp. "Rubber Ducky"
Thailand

(Order Glomerida)
Rhopalomeris carnifex
Thailand

(Order Spirobolida)
Narceus americanus albino cultivar
Ozopores are visible as dark marks along the side.

(Order Spirobolida)
Narceus americanus albino cultivar
The first instar has one set (2) of ozopores, while the second instar has sixteen sets (32).

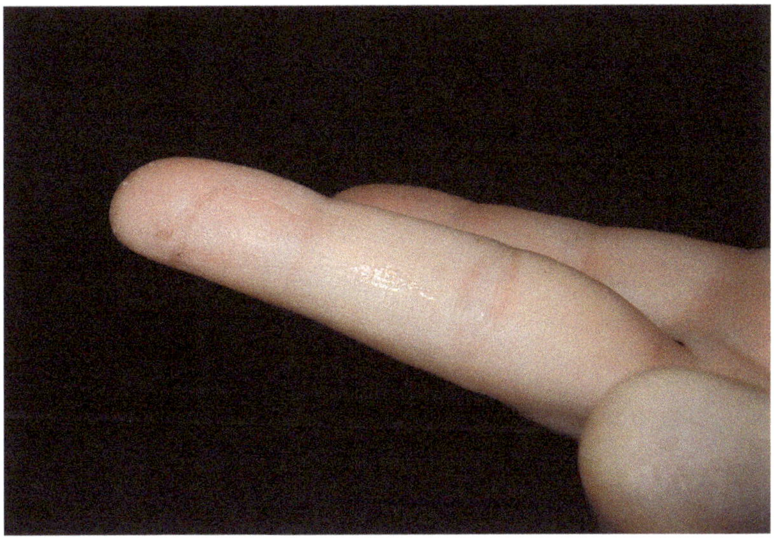

Glomerin exudate spread on middle section of finger

(Order Spirobolida)
Centrobolus sp.; Mozambique
Ozopores are often brightly marked on young millipeds.

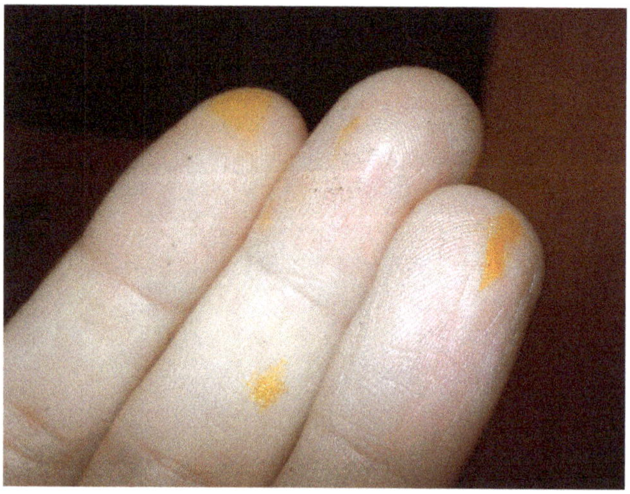

(Top) (Order Spirobolida)
Defensive fluid can be seen oozing from the ozopores of this *Acladocricus* sp. "Philippine Giant Blue." Ozopores are located on most segments on both sides of the body, but only those near where the animal was touched were employed.

(Middle) *Acladocricus* defensive fluid fresh on skin
The secretion is yellow on first contact. The skin did not turn red-brown because it was washed off in less than thirty seconds. A spray bottle was kept nearby to rinse off exudate quickly.

(Bottom) (Order Spirobolida)
Floridobolus floydi
Secretion from nearly every ozopore

Feeding begins as they chew under the milliped's face. A full-grown milliped hunter is often much wider than the milliped at rest, but it can telescope part of its body all the way through the prey as it meticulously cleans out the entire milliped and leaves the empty shell behind. One of the largest species, *Zarhipis integripennis,* from the southwestern U.S., is 3″ (7.6 cm) in normal resting position but can stretch the full length of the longest *Orthoporus* (>6″ or 15.2 cm) as it feeds on its natural prey. It is brilliantly colored in alternating orange and black segments. Adult female specimens live for years in captivity and are not harmed by waiting months between meals. The most amazing characteristic of these bizarre milliped specialists is the pattern of glowing, concentric bands that run the length of the body. Members of this beetle family are more commonly known as glowworms than milliped hunters.

Teratology

Rare congenital defects are most often seen as conjoined segments at a rate of one out of thousands. Gynandromorphs (half male and half female) probably occur at a similar rate but may not be very noticeable. There seems to be only one report of a gynandromorph specimen by Carl (1932) of *Coonoorophilus monstruosus*, a Sri Lankan polydesmidan. Conjoined segments have been reported for individual specimens of four different spirobolidans: *Chicobolus spinigerus, Atopochetus dollfusi, Apeuthes* sp. and *Narceus americanus* (McMonigle 2012, 2021). All of these except the *Atopochetus* were captive-bred specimens.

Over the years there have been various reports of pigment disorders, specifically albinism (amelanism) in arthropods, but verifiable reports are extremely rare outside of crustacea. Even cases of incomplete albinism (such as a pied *Blaberus giganteus*

This millipede hunter, or glowworm, from Texas is unimpressed by the hard exoskeltons and chemical defenses of a milliped.

(Order Spirobolida)
Narceus americanus albino cultivar

(McMonigle 2022b)) and erythrism (pink, yellow, or orange katydids cultured at the defunct Audubon Insectarium, New Orleans) are relatively rare for insects. A few albinism reports have been published in journals (Putnam 1958; Locket 1986; Yerushalmi et al. 2001). The majority of popular reports of invertebrate albinism on the internet or in articles refer to pale forms (hypomelanism) often within standard variation, freshly molted animals, or species such as those found in caves that are naturally white in color. Albinism should only refer to an aberrant form of an animal different from the standard coloration because specimens lack the ability to produce melanin (or other primary dark pigment). For cave animals this would only hold true if the troglobitic line fully lacked the capacity to produce melanin and a naturally colored form of the same species could be found outside the cave.

True albinism documented for milliped species is so far limited to a single stock of cultivars of *Narceus americanus* (Spirobolida) isolated in captivity between 2002-2004 (McMonigle 2008) and cultured specimens of *Ophyiulus pilosus* (Julida)(McMonigle 2014) that were not isolated into a pure-breeding stock. Considering both documented cases of albinism in millipeds are recent and by the same author, it seems likely the extreme paucity of diplopod albinism reports reflects more on the rarity of milliped husbandry and research than on the scarcity of the amelanistic pigment disorder in this class of arthropods.

(Order Julida) Adult *Ophyiulus pilosus*, one albino

(Order Spirostreptida)
Orthoporus sp.
Erythristic?
Val Verde Co., Texas
Pufferchung (CC BY 4.0)

(Order Polydesmida)
Apheloria virginiensis
Congenital segment deformity

(Order Spirobolida)
Atopochetus dollfussi
Vietnam
Conjoined segments

(Order Spirobolida)
Narceus americanus albino cultivar
Conjoined segments

(Order Polydesmida) *Anoplodesmus saussurii*; Kerala, India
Charles Sharp (CC BY-SA 4.0)

(Order Spirostreptida) *Eumastigonus* sp.; New Zealand
Shaun Lee (CC BY 4.0)

HUSBANDRY

Artificial Habitats

Housing for large members of the Diplopoda is only effectively judged long-term. A horribly inadequate habitat can still take many months to kill the animals. Millipeds can die very slowly, hanging on for months after they are beyond hope of recovery. As mentioned, an imported *A. gigas* with a severely damaged head that could not feed lived four months following its import in 1998. Feather millipeds and small pill millipeds (glomeridans) can survive six months without the correct food. The conditions outlined in this chapter have been used to rear tens of thousands of young from dozens of species (mostly Order Spirobolida) and to keep adults alive for as long as fourteen years.

The chosen container should prevent escape. If the animal gets loose and dries up in the corner of the facility, weeks of painstaking exactitude in cage preparation and design are wasted and the animal is dead. Avenues for escape should be carefully considered and avoided. Ventilation holes should be covered with screen if there is a chance the millipeds can get the head through. Millipeds get stuck partway in drilled holes and the enclosure will have to be carefully cut apart to attempt to remove the animal without tearing it in half. Although diplopods (other than polyxenids) cannot climb smooth, upright surfaces, the walls of enclosures often end up with substrate or frass stuck to them, or as with polydesmidans, the frass is habitually pasted up the sides. Many large, round millipeds are arboreal and climb the corners of glass terraria even if the silicone bead is removed. Plastic shoeboxes should have lids that snap securely given that container sides are normally at a slight angle, and the distance from the substrate to the lid is short. Care must also be taken to ensure the lid is secured following every use. The perfect enclosure is worthless on the day the lid is left ajar.

Moisture and ventilation are important variables leading to long life or rapid death. Millipeds do burrow to escape desiccation, but with a constantly dry surface, the animals stress and die prematurely. Although millipeds normally hide, with low moisture they occasionally surface to race around the cage in hopes of escaping the nightmarishly dry habitat. If the moisture is correct, they are often found on or near the surface at night (except during molting and while laying eggs). The less ventilation there is, the easier it is to maintain high moisture. On the other hand, if ventilation is inadequate, the animal can become stressed and eventually die. Furthermore, minimal ventilation and high humidity along with overfeeding dog food or fruit can poison the milliped's environment as the food rots.

Several factors are considered in controlling and determining moisture levels. Ambient room humidity affects the required ventilation. Humidity levels in human structures tend to be dry, but are still extremely variable by the season. The captive habitat's substrate may need to be watered every week in the winter, but not at all during the humid springtime or rainy season. Airflow is equally significant because high flow rapidly dries substrate despite high humidity. In addition, evaporation increases as the temperature goes up. Due to external variables, a digital readout of the enclosure's humidity has limited practical application (though below 35% humidity should be avoided). Practical care involves use of limited ventilation, paying attention to the speed at which the substrate dries, and the health of the animals. The surface of damp substrate should not dry out in less than a week. If room humidity is high and temperature and airflow are low, the cage may not require a lid. If the substrate's surface dries out in a few days or less, the ventilation needs to be drastically reduced. It does not matter if the bottom layer of the substrate is damp, because the millipeds can still desiccate. Another moisture consideration is related to slight temperature differences on the outer walls of the habitat. Even with limited ventilation, over time one side or area can become bone dry as the moisture concentrates on the cooler side. If this happens the enclosure or substrate should be rotated. Very young specimens and tiny species and can be greatly harmed by extreme moisture gradients.

Suggested Ventilation for Large-Bodied Millipeds

This will be useful under most indoor conditions; reduce ventilation by half if room humidity is below 20%.

1. Plastic shoeboxes commonly have loose-fitting lids that provide adequate ventilation. If the lid seal appears airtight, eight to ten 0.0625 " (1.6 mm) holes should be drilled around the top of the lid. Position holes so they will not be covered up when containers are stacked. A larger number of thumbtack holes can be used if housing millipeds small enough to squeeze through drilled holes.

2. Watertight glass terraria with a glass lid can have a 0.25 " (6 mm) gap left open along one side (if the species contained is not able to scale the corners). The inevitable gaps around the commonly available glass hoods also offer an acceptable level of ventilation. If the substrate depth is less than three inches, the air space together with the usual gaps under a solid glass lid can provide adequate airflow.

Millipeds live indefinitely without ever drinking water as long as the substrate is damp and ventilation and airflow are not excessive. If they come running to a water dish there is a problem. Arthropods have a very different cellular waste removal system than vertebrates and do not require 'flush' water for ammonia removal. If the milliped is thirsty the cage is too dry. A drinking water dish is not a necessary part of a good milliped habitat. However, severely dehydrated millipeds may be found with their rear ends in the water dish as they can rapidly uptake water through the back end.

(Order Spirobolida)
Climbing *Apeuthes*
Glass terrarium's top edge is 12" (30.5 cm) from bottom.

Tropical and temperate millipeds are normally kept near room temperature and do well anywhere from 65°-82° Fahrenheit (18°-28° Celsius), though some species prefer a narrower range from 73°-78° F (23°-26° C). Native North American millipeds, especially *Brachycybe, Californiulus, Euryurus, Harpaphe, Narceus,* and *Tylobolus,* prefer temperatures below 75° F (24° C), but will not move much nor eat if it drops closer to 50° F (10° C). Even species native to sweltering hot, tropical countries can experience die-off if kept above 80° F (27° C) for long periods. Tropicals are often accustomed to much cooler understory or underground microhabitats, and only *Orthoporus* is ever caught basking in the sun on a rock. High temperature of 90° F (32° C) or so for a few days does not kill most common species, but can quickly eradicate most flat millipeds and northwestern round millipeds.

Decorations in artificial enclosures can be anything the habitat creator desires. Tufa rocks, ceramic castles, green marbles, plastic corals, and tiny garden gnomes are acceptable. A large rotten log is suggested because it looks natural and provides food (it is important to remember to heat treat the log to kill unwanted guests). Individual decorations heavier than two ounces (>57 grams) should be evaluated in cages where numerous immatures are present. Molting animals prefer to form cells below large objects and tank mates enjoy digging under the same object. A heavy rock or log becomes a very efficient molting-milliped-smashing-machine when there is a high population density.

Living plants can be planted in milliped habitats if the population density is reasonable (no more than half a dozen large millipeds in a 10-gallon/38-liter). Millipeds usually will not eat live terrarium plants like parlor palm, silver inch plant, Pothos, ivy, croton, vine fig, etc. The reason for a low population density is that numerous burrowing animals in a small cage will unearth the plants. Remember that store-bought houseplants may be treated with systemic pesticides that could kill millipeds for up to a year, if they eat the soil. Wait several months before introduction so the pesticides can break down, or purchase plants from a source that guarantees it does not use persistent pesticides. Organic pesticides can be just as deadly to millipeds as synthetic pesticides.

Fluorescent bulbs or LEDs can be used to light the terrarium and keep the plants alive and growing. Pothos, English ivy, and silver inch plant do well with reasonable ambient room lighting, but many plants require additional light. Incandescent lighting placed directly on the cage, and sunlight, should be avoided due to overheating and dangerous moisture gradients. Habitats without plants require no lighting. A day-night cycle is unnecessary, but is still provided by indirect lighting unless the caging is carefully shielded from light leaks.

Enclosure dimensions should be, at minimum, twice the length of the largest specimen. Large cages can be difficult to clean and replace substrate. The number of specimens that coexist safely in one enclosure has many factors beyond the size and type of milliped. Up to two dozen adult *Narceus americanus* can do well in a 10-gallon/38-liter enclosure with about 20 liters of wood-based substrate—or six adult AGBs or hundreds of *Spirobollelus*. As they are mostly subterranean, substrate depth equal or greater to adult length is suggested. Overcrowding is relative to various aspects

(Order Spirostreptida)
Captive-reared *Archispirostreptus gigas* mating while hanging from large branches in terrarium

of the habitat including substrate depth and composition, size, arrangement, and number of cage decorations, and the level of care provided.

One aspect of milliped caging, which appeals to display design, is that a variety of species can be safely maintained in a community terrarium. A display can be designed to represent invertebrate groups from a specific biotope or entail a colorful variety of diplopods like a community fish tank. Terrestrial isopods (types that do not reproduce rapidly), humid forest phasmids (esp. members of the Heteropteryginae), cockroaches, land snails, cave crickets, and other invertebrates with similar habitat requirements will safely coexist with millipeds. Furthermore, most of the various sizes, shapes and colors of milliped have similar requirements and can be kept together with very little risk of injury. The drawback to keeping different species together is when babies appear, it takes six months or more to tell whose is whose. Community habitats can be easily maintained and create enjoyable displays, but it is important to harvest creatures that reproduce rapidly before overcrowding damages all the inhabitants. The most visually impressive invertebrate display terrarium I have encountered was a 55-gallon (208 L, 48″ x 13″ x 21″) aquarium containing at least a hundred *Archispirostreptus gigas* (captive produced from a few wild adults previously added to the display) and more than a hundred large South American *Blaberus giganteus* visible throughout upright logs and branches that stretched the 48″ (122 cm) width and 21″ (53 cm) height of the display (Cleveland Botanical Garden circa 2006).

Habitats should be monitored for moisture levels and substrate quality. If substrate dries it should be watered but be careful not to add too much water. (Water may be added unwittingly with certain foods, especially watermelon.) Substrate should never be wet enough to create anaerobic conditions. Edible substrate might look okay but no longer inspire feeding behavior in the animals. Add a few handfuls of substrate when pieces of wood or leaves in the cage no longer inspire interest. Replace the substrate if pests proliferate or there is a large proportion of frass. Check frass for the presence of egg capsules before discarding. (If not a single hatchling or egg capsule is revealed, there is a very good chance there are none; but if a single larva or egg is seen there are likely many more.) Substrate can be stored in a sealed plastic bag and checked for larvae after a few months. Do not throw used substrate out in the yard unless you are only keeping species collected from your local area. Used substrate can be thoroughly cooked at high temperatures, placed in a plastic bag, and discarded in the trash.

(Order Spirobolida)
Narceus annularis

(Order Polydesmida)
Euryurus maculatus
Florida
Polydesmids like to paste frass all over the sides of their container and can use that to climb out.

(Order Spirobolida)
Anadenobolus modestior
2" (5 cm) at ten months, in a rotten wood substrate

Substrate Materials and Pests

Millipeds eat whatever material surrounds them, so anything used as substrate also becomes food. For this reason, vermiculite, sand, perlite and other materials without nutritional value should, at most, make up a small fraction of the substrate. An endless list of substances has been used for milliped bedding, but some common substances like gravel and wood chips provide an unhealthy environment for the animals.

Habitat substrate should include a base and a food component. A base substrate is used for the bottom inch or two, and edible substrate is placed on top. The base retains moisture and allows for burrowing. A portion of the food substrate is mixed in so that very young millipeds do not have to travel far to feed. The three food substrates listed are often used together with a ratio that accords to individual species requirements. Alternatives for the base and food are listed:

BASE SUBSTRATE

Compost: Ingredients vary greatly by brand, region, and even batch. The main ingredient in mushroom compost is manure from herbivores (cow, horse, chicken), while the main ingredient in compost manure is often yard waste, but they usually all mostly contain decomposing vegetation, the natural diet of the diplopod. Assuming nutritive value is based on the growth and health of immature millipeds, composts are the best base. I have used Garden Magic composted manure for decades with good success. While some bags have rocks, sticks, and plastic waste mixed in in quantity, it still works. The main ingredients are supposed to be decayed hypnum moss and cow manure. Other composted manure brands tested short-term also seemed to work. Mushroom compost is a similar, nutritive base that is usually made of decomposing straw, horse bedding, poultry manure, and corn cob. Unfortunately, the quality and components are not consistent, which is worrisome when each unique batch is only testable on live animals. Composts are the most likely base to become infested with fungus gnats and are commonly infested with soil nematodes. Since the main ingredients are agricultural waste, there is always a chance of introducing predatory nematodes, pathogens, or pesticides intended for the purpose of arthropod control.

Potting soil: This is the easiest base to find at almost any corner or department store, and it had been the most commonly used base prior to captive breeding. Ingredients vary greatly and include sand, peat moss, pine bark, composted manure, perlite, lime, vermiculite, and various substances that may or may not be healthy for millipeds. Potting soil tends to lack decaying vegetation as a primary ingredient, but some have more manure than a compost.

Raised bed potting soils are a better choice since they contain a larger portion of composted material. As a mix of compost and inorganic materials, potting soil presents the greatest chance of lacking adequate nutrition and a reduced chance of fostering pests.

Hardwood pellets: These are expanded in water until they attain a soil-like consistency. A number of different mixed hardwood brands can be used with good success. The types of trees and ratios are not usually listed, and pure oak or cottonwood may work better or more consistently than a mix. Hardwood is better than coconut fiber since it does not have stringy fibers and it will eventually break down into something edible. Still, some wood types may be harmful. Common grill flavors like mesquite, apple, and cherry have some insecticidal compounds which may require a period of decomposition before use. Initially there is close to zero nutritional value and diplopods will starve like a human would if only fed sawdust. After the wood sits wet for many months and is colonized by fungi, or if the decomposition process is sped up artificially, it becomes a primary food substrate (rotten wood).

Coconut fiber or coir: This is not a good food base like most composts and many potting soils, but it has consistent qualities despite brand and region. If the additive food types and sources have been perfected, there is relatively no chance of an accidental introduction of pathogens, pests, or poisonous materials. Results are more consistent and coconut fiber is commonly available for purchase as reptile bedding and orchid medium. It is available in compressed bricks under a variety of brand names. The finely ground type is used (large chunks are useless). One useful quality is its color changes with the presences or absence of water. Color change makes monitoring for moisture extremely easy since it is light orange brown when it dry, and dark brown when wet.

The main problem is that coconut fiber has almost no nutritional value, so decayed wood, leaves, and supplements have to be added to provide adequate nutrition or the animals starve. Coir is commonly used in mushroom substrates, but the unprocessed material has low nutritional value like fresh wood flakes. Also, coir contains stringy fibers that are part of the coconut shell. Since millipeds eat the substrate, and the fibers pass through the digestive tract and stick out of the frass, some have recently suggested the stringy fibers cause blockage in the digestive tract. For decades breeders used coir as a component successfully, but it has become controversial in the last year or two.

FOOD SUBSTRATES

Rotten Wood: This is an extremely variable medium given not only the number of different species of pines and hardwoods, but also the species of associated fungi and level of decay. Identification of nearby trees may help determine the type of rotting log found in the woods. No matter how rotted, perpendicular branching usually makes pine very easy to discern from hardwood. Rotten hardwood is suggested. Softwood (conifers) like pine should be avoided, but can provide safe nutrition when it is extremely decayed. No matter what species of tree is chosen, the wood must be soft enough to be easily scraped with a fingernail, otherwise milliped mouthparts will not be able to bite into it. The best wood is decomposed enough to be easily broken and crushed with bare hands. In nature, various invertebrates including julidans, terrestrial isopods, and beetle grubs normally feed on rotten wood. If no invertebrates appear to be eating any part of a piece of wood, that piece should be avoided. Brown to dark brown rotten wood is usually the most edible, but extremely dark brown to black wood is generally too decayed to retain nutritional value.

Small, decaying branches that fall off trees are usually a good source of rotten wood, often best when covered in lichen. Many millipeds also eat the lichen, but feeding is primarily dependent on the type of tree and level of decay of the wood rather than the type of lichen. Branches from black locust (*Robinia*) and coffee tree (*Gymnocladus*) are loved by various millipeds.

Rotten Leaves: Generally only angiosperm (flowering plant) leaves are used, though a few pine needles mixed in is normally harmless and ignored. Aromatic hardwood leaves like eucalyptus and any conifer needles must be extremely decayed to be acceptable as food. A mix of leaves works well, but at least one of these: oak, beech, elm, locust, *Viburnum*, or rose, should be part of the mix for best results. Only leaves that have been on the ground decomposing for at least a few months should be collected. If there is no other option, leaves taken off the plant or freshly fallen can be collected, crushed, and baked. However, recently green leaves can be extremely volatile and should be used sparingly to prevent grain mite infestations. Leaves should be finely ground to facilitate decomposition and feeding. Whole leaves can sit on the surface of the substrate for months and may not even be nibbled at the edges. The same leaves ground into tiny flakes or powered can be eaten in volume within days. Large pieces may end up cage decoration rather than food.

Shredded Wood Bedding or Sawdust: This is better mixed into the substrate rather than spread on top. Unless sawdust is very finely ground, it does not have the consistency for a base substrate. Fresh sawdust is almost entirely inedible but the surface area to mass allows it to be rapidly colonized by fungus (which is nutritious) when kept damp and buried or mixed with milliped frass. Millipeds can start nibbling on it within weeks when mixed with used substrate, though a few months is more likely. Poplar bedding, aspen bedding, and mixed hardwood chainsaw dust have been successfully used. Softwood sawdust or bedding such as cedar and pine should be avoided. Ground up wood is a secondary replacement for rotten wood. Since the species of tree can be known, it may be more consistently applied. This material is unlikely to be a source of harmful, introduced life forms, but it may need to decompose for long periods before any portion is edible.

Substrate materials can be stored safely for decades. Unused leaves, wood, and sawdust should be stored dry. Damp or wet storage can degrade nutritional value. The primary problem with damp storage is not fungus. Rather, pest infestations can reduce the material to a pile of useless worm or maggot frass.

All substrate materials and decorations collected from outdoors should be thoroughly cooked at a low temperature 150-200° F (65-93° C) to kill pests. (Freezing cannot kill most pests of materials collected in temperate environments.) Higher temperatures are not necessary and may damage the materials. If wood or leaves are singed or charred, those sections are usually inedible. If substrate is placed in a thin layer on a shallow tray, the time required for thorough heating is twenty to thirty minutes. If it is in a deeper roaster, a few hours may be necessary for thorough heating.

Seemingly lifeless materials collected from outdoors can conceal an army of pests. Nematodes, earthworms, julidans, isopods, snails, earwigs, and mites should be prevented from invading the captive habitat. Adult giant millipeds are not directly damaged by any of these pests, but certain pests destroy the food value of the substrate. Avoiding pests is extremely important for raising very young millipeds. The egg stage through the first few molts is extremely vulnerable and any pest can damage or kill them. Once the millipeds have molted a few times, most pests cannot easily harm them. However, heavy infestations of earthworms, fungus gnat maggots, and large roundworms can be deadly. Young millipeds starve if food is not nearby and will be completely wiped out if worms have already eaten much of the substrate. A small percentage of spirostreptidan young can survive heavy pest infestations, while spirobolidan babies die out completely with moderate pest problems. For the hundreds of times I have wished I was more careful preventing pests through proper cooking of all materials, there is not a single time I regretted cooking the substrate (though I never accidentally cooked any millipeds nor set my oven on fire).

Keep in mind that even if pests have not overrun the substrate, it is necessary to replace or add to the substrate from time to time. Rotten leaves and rotten wood that remain uneaten for long periods have not warranted the interest of the millipeds, so new leaves or wood should be added. Expanded wood pellets and saw dust can be mixed into old substrate to extend its usefulness.

PEST PROFILES

Julidan millipeds, land snails, slugs, and terrestrial isopods eat the food and substrate, but are not a major concern. However, they tend to breed much more rapidly than large-bodied millipeds and thousands of them are not harmless. Massive isopod infestations will consume eggs, larvae, and even freshly molted isopods that are exposed.

Earwigs can eat tiny millipeds and eggs, but they are unlikely to occur in numbers that could make an observable dent in a normal size milliped brood.

Nematodes are tiny white roundworms (less than 0.25″ (0.6 cm) long and the thickness of a human hair) that destroy the food value of substrate and kill eggs and early stage millipeds. These are troublesome pests because they are difficult to prevent, reproduce in plague proportions, and still can go unnoticed because they are so tiny. They are prevalent in most dirt, soil, rotten wood, and leaves. Attempting to eradicate nematodes from an infested enclosure will prove fruitless if one speck of old substrate is retained. They even hide beneath loose silicone in reused aquariums so extra measures such as an extended dry period or bleach soak is needed. Though difficult to eradicate, it takes them a while to completely ruin the substrate (especially wood).

Dark-winged fungus gnats (Diptera: Sciaridae) are small flies that can infest most types of substrate, but larger species only reproduce in plague proportions on rotting leaves and some types of decayed wood. They commonly infest the soil of potted plants and are a bane to mushroom growers. Fungus gnats are found outdoors worldwide and find their way into human structures because they fit through window screen. Sometimes entire garbage bags of leaves can be turned into useless maggot frass. Other times the flies are just annoying since they like to fly in human eyes and noses. Enclosure ventilation should be covered with microscreen or provided by thin pin holes (small thumb tack holes are too large) to prevent infestations. Larger types are the length of a small fruit fly, but thinner. There is a much smaller species that reproduces in extreme boom-and-bust cycles and is more difficult to keep out of containers even with guarded ventilation.

Earthworms are a milliped zoologist's nightmare because they quickly turn substrate into worm feces and can go undetected for a long time despite their large size. These large oligochaete worms devour dead leaves and rotten wood, thoroughly destroying substrate. Earthworms can be accidentally introduced with untreated substrate materials, especially rotten wood, soil, leaves, and moss. Even mature millipeds start dying when the substate is mostly worm castings. Although earthworms are large, often 1-3″ (2.5-7.6 cm) long, they are virtually invisible since their slime allows them to wear a coat of surrounding materials. Earthworms do not eat small, living millipeds, but they collapse molting cells, often resulting in death. If earthworms heavily infest the habitat of a milliped that molts concurrently, every last milliped may be killed before the problem is noticed. Earthworm eggs look similar to giant milliped eggs, but have much tougher shells, are never found in egg capsules, and may have a tiny stem on the side.

Grain mites are round and do not jump. They are extremely tiny, white or cream-colored arachnids from the Order Acari. They are semi-opaque, so the gut content can make them temporarily the color of the foods they infest. The commonly encountered are members of the Family Acaridae, known as storage mites. The common name comes from their habit of infesting grain and food products. In milliped habitats, grain mite infestations are caused by overfeeding supplemental food, volatile composts, and use of recently fallen or fresh leaves. They can also proliferate after pest-infested substrate has been heat treated, as they reinfest quickly and feed on the remains of pests including their own. In rancid conditions they reproduce in plague proportions and are associated with dying millipeds. Infestations are the direct result of poorly prepared substrate and overfeeding—the real reason the millipeds are sick. If a mite outbreak occurs, replace the substrate and clean the cage. Large outbreaks should be treated with complete substrate replacement and can be followed by use of only old rotten wood and well-decayed leaves. Large infestations give the cage a sickening-sweet smell like applesauce mixed with wet dog food. Cooking substrate after a large infestation is a very temporary fix because mites jump cages (it is not easy to safely remove every mite from the millipeds) and repopulate to become even worse. The most common grain mites are the tiny, white specks that grow out of control with overfeeding. At full size they are smaller than a grain of salt and a magnifier is needed to make out the head and legs. Tiny grain mites form a crust on the milliped's body surfaces and are a sign of spoiled substrate. They do not feed directly on the milliped but cling to the exoskeleton and may stress or suffocate millipeds in extreme cases. If the milliped's body becomes coated with large patches of mites following a bad infestation, mites can

(Order Polydesmida)
Oxidus gracilis can be a pest outdoors, but unlike julidans, it is easy to manually remove from cages.

(Order Julida)
Common pest julidan millipeds are often impossible to remove without a complete substrate change. A few are always missed and they grow back quickly.

be removed with a cotton swab and alcohol. Care must be taken to avoid the spiracles and head. It is possible to damage or kill the milliped, so it safest to improve conditions and wait for the mites to die out. Pest mites will eventually wither away with improved conditions. Grain mites are everywhere, and some are inevitable. The key is to prevent large outbreaks by feeding the millipeds optimally.

Most mites are not troublesome habitat pests. Commonly encountered free-living mites grow slightly larger than grain mites and tend to be black or red in color. These infest uneaten food, but fortunately they do not cling to the milliped following infestations. Predatory mites are even larger (still extremely tiny but big enough their legs can be seen with the naked eye). They are fast moving and commonly tan in color. Predatory species feed on grain mites and reduce infestations over time, but will not react to grain mite infestations quickly nor counter bad conditions. Commensal mites (addressed in the biology chapter) are by far the largest mites associated with millipeds, but they are harmless and unaffected by overfeeding. A fourth grouping of mite (parasitic) has not been encountered on captive giant millipeds.

Fruit flies, *Drosophila* species, are encountered in habitats from time to time when feeding with fruit. They are not a significant pest since the fix is to simply not leave uneaten, rotting fruit in the cage—it should not be there anyway.

Phorid flies may also be encountered but they are simply attracted to dead arthropods and are the vultures of the invertebrate world (if vultures, instead of feeding, laid hundreds of eggs on the dead creatures which rapidly developed into maggots that consumed the carrion). The Phoridae are also known as 'scuttle flies' because they walk along surfaces as much as fly.

Springtails are tiny, harmless hexapods and the only pest that does not bother even the most sensitive spirobolidan babies. More than one fledgling zoologist has mistaken springtails for mites. However, springtails jump, move rapidly, and most have long, thin bodies. Mites do not spring into the air. Under magnification they look very different from arachnids because springtails have three body parts, only six legs, and a pair of antennae. Many types are commonly introduced with untreated soil, leaves, wood, and moss. Most are white or silver and incredibly tiny. Some are a little larger and very long, like *Folsomia candida.* A few bright orange, yellow, or blue species are cultured for clean-up crews. Almost any species can have some positive impact on reducing grain mite populations through food consumption. The one species recommended for competing with mites is the popular and prolific 'large white' AKA 'pink tropical springtail' commonly available from vendors. It is successfully used to reduce mite populations through food competition. Older specimens only grow as large as a very small fruit fly. If you look closely, you will find few images in this book devoid of this specific springtail species. It works very well, but only for leftover particles and minor overfeeding (as nothing can counter rancid conditions). This species was specifically selected and bred starting in 1992 for the purpose of reducing grain mite problems when keeping giant millipeds.

Grain mites have three primary behavioral phases: reproductive, wandering, and dispersal. This photo shows hundreds of reproductive mites feeding on a small piece of puppy food.

If wandering grain mites do not find food and space, they undergo a molt resulting in short legs and a disk-like body (hypopus stage). This dispersal form allows them to cling tightly to the skeletons of large arthropods, like millipeds, and catch a ride. The shape protects them from desiccation and removal. Unless the wandering stage is severe (as seen in this infestation), zoologists may not notice the problem before mites have developed into the hypopus stage.

This terrestrial snail survived ninety days of drying. Very few other pests can withstand long-term drying.

Mold is usually the milliped culturist's friend because it slowly transforms leaves and wood into edible substrate. The most common unwanted types are various molds that attack fruit and dog food and render it inedible. However, food should not sit around long enough to be completely enveloped in mold, so it may be a positive alternative to maggots and mites. There is one problematic pest mold which is common. It is slow growing and requires months to destroy a captive habitat. Its presence does not rapidly harm millipeds, but if left unchecked for the 4-12 months it takes to engulf a habitat, then all the millipeds in the cage can die. The mold itself is composed of unassuming white strands and puffs and does not produce mushrooms. It is easy to differentiate from other common white molds because it changes the infected substrate color from black or brown to a light yellow. It also has a strong smell. This fungus is usually introduced with rotting leaves but will also affect wood mixed with the leaves. It usually only grows at 74° F (23° C) or higher temperatures, and when substrate is on the dry side of damp. Growth can stop with high moisture levels. Since it grows slowly, pieces can be removed when they show up (remember most white molds are harmless), but it is usually less work to clean the habitat and replace the substrate.

Supplemental Foods

Millipeds can survive on substrate alone, but reproduce and grow faster and larger with the additional foods listed here. These foods are highly volatile and lead to pest outbreaks, so it is important to avoid overfeeding. For most species a small amount can be offered every week or two. (Some species should not be fed supplemental foods at all or fed only as a rare treat.) Uneaten portions of foods that quickly spoil should be removed after 72 hours. The frequency of feeding should be cut back if food is not consumed with zeal. Dishes to hold the food can be used with large animals, but small specimens will not be able to get to the food. Dishes are minimally practical as millipeds often pick up and relocate the contents.

Feeding voracity can be used to determine the usefulness of the food. Animals should be found feeding on the food within minutes to hours, though feeding zeal can be seasonal or related to where they are in the growth cycle. (Some immature millipeds, primarily spirostreptidans, synchronize molting when kept in the same cage even if they are at different stages and levels of growth.) Healthy animals tend to have the heartiest appetites and species that rarely eat foods other than leaves and wood may feed on other items at times. Species that relish supplemental foods may totally ignore them if rotten leaves or wood are available in sufficient quantity and quality. Foods that are usually ignored by a species may attract attention when the enclosure is overpopulated.

Fruits and Vegetables: Fresh melon rinds, pumpkin, and squash are useful foods for a wide variety of millipeds. Items which inspire the greatest interest from the widest variety of millipeds are the rinds of pumpkin, cantaloupe, and watermelon. Pumpkin is seasonally available in September and October and can stay fresh through March if uncut (though it is possible to store mini pumpkins for a year). Pumpkins can be very inexpensive or free depending on your costume and how fast you can run. Cantaloupe is always available and is of greater interest to the animals, but it molds quickly and can lead to excessive substrate moisture over time. Watermelon rind is the favorite of many millipeds, but it is somewhat seasonal and really does live up to its name: consistent use can lead to extremely waterlogged substrate. Raw carrots and potatoes may stay in the enclosure for months without rotting, but they also rarely inspire feeding. Romaine lettuce is of great interest to some of the large juliforms. Depending on freshness, if part of the crown is included, and enclosure conditions (light and moisture), romaine can last months until eaten or it may melt into black slime in a few days. Leaf and iceberg lettuce should be avoided because they offer minimal nutrients—few, if any, millipeds show interest in raw spinach. Apple and sweet potato slices are slow to spoil, but many millipeds grow bored of them after a few nibbles. Banana slices can be used but are usually ignored. Some vegetables including carrots, green beans, and sweet corn are mostly ignored when raw but will be eaten if cooked. Uneaten leftovers from cooked vegetables should be removed within a few days to prevent mold growth.

Lichens and Nonvascular Plants: Larger lichens and mosses whose bracts and leaf-like parts are easily visible to the human eye are mostly inedible. Large branchy and leafy lichens are refused almost universally. Large moss like sphagnum and peat are refused (unless ground and mixed into edible substrate) and lack nutrients. Crustose lichen that grows on the stems of woody plants is eaten by many species, though the value is uncertain, and feeding is mostly determined by the species and decay level of the bark beneath. Micro mosses whose bracts (leaves) are barely discernible with the naked eye are fed on by various millipeds. However, feeding is concentrated on the dirt and materials held together by the rhizoids below the living portion of the moss and seems to be determined by the quantity of ground algae (*Nostoc* spp. or trolls' butter) more than the moss or detritus. Liverworts are untouched unless dead and decayed.

Algae: In addition to lichen and the blue-green algae (cyanobacteria) mentioned above, millipeds will eat red, brown, and green algae. Freshwater hair algae (green) is eaten by many species but not with great zeal. Black beard algae (red), *Audouinella* sp., is of some interest to glomeridans and polydesmidans. Prophase kelp (brown) from the local Asian grocery store can be washed of salt—a couple of the big Juliformia will eat it. Pelleted fish food for herbivores made with algae meal is an attractive food for many different species (eaten at about the same rate as dog food). However, herbivore pellets are not recommended—while they do seem to improve growth of some species, this food results in massive molting deformities at a rate of 80-100% for certain species including *Archispirostreptus gigas* and *Trigoniulus macropygus*. However, it did not cause any molting problems for *Acladocricus* ex. Negros Island or *Narceus americanus*.

Kibbled Pet Foods: This includes dog food, cat food, and fish food pellets or flakes. The exact brand or flavor does not seem to make a big difference—monkey chow and ferret pellets also work. Pelleted food for rodents, mostly made from hay and grasses, molds readily and is inedible unless heavily decayed. Fish food pellets for predatory fish work well, but algae disks and pellets with high levels of spirulina lead to molting problems for some species. The most useful are semi-soft pellets (usually sold as cat or dog treats or included as soft kibbles in a food mix, not the bits), because they do not mold as quickly. Large pieces of food should not be smashed or ground up. If the millipeds will not feed on the whole piece they will not feed on ground up pieces and removal of unconsumed particles is difficult. Fish food flakes offer the illusion of successful feeding, but only because they are thin and appear to be eaten rapidly. The problem is fish flakes often dissolve into damp substrate and most or all can be wasted.

Powdered Foods: This includes whey or soy protein isolate, powdered milk, and baby formula. These are effective foods used in place of dry pet foods. A pinch or two is placed on the substrate and then misted until wet. Despite the very different nutrients involved, they seem to work equally well in aiding development and are eaten with similar interest. Feeding should be

This colony of *Apeuthes* sp. resides with orange *Porcellio scaber* isopods. There are thousands of isopods in this cage that damage and consume the millipeds from time to time, but mostly coexist without issue.

(Order Spirostreptida)
Choctella cumminsi
Feeding on dog food

(Order Polydesmida)
Orthomorpha coarctata
Immatures eating dog food

(Order Polydesmida)
Orthomorpha coarctata
Feeding on *Audouinella* sp. red algae

(Order Glomerida)
Glomeris pustulata
Young eating pumpkin

(Order Spirostreptida)
Cambala annulata
Dog food is well received.

(Order Julida)
Californiulus yosemitensis
Feeding on dog food

(Order Spirobolida)
Narceus americanus albino cultivar
Feeding on watermelon

(Order Spirostreptida)
Spirostreptus sp.
Young feeding on pumpkin

(Order Spirobolida)
Spirobollelus sp.
Feeding on pumpkin

(Order Spirobolida)
Leptogoniulus sorornus
Feeding on decayed black locust wood

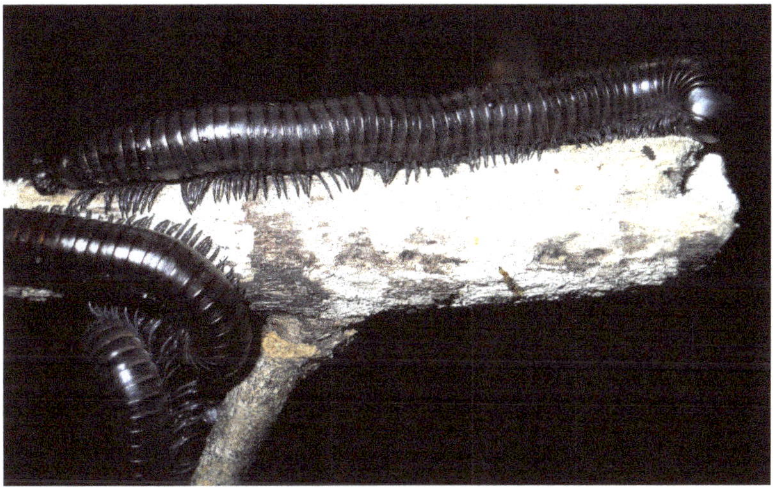

(Order Spirobolida)
Anadenobolus arboreus
Virgin Islands
Feeding on decayed black locust wood

(Order Spirobolida)
Apeuthes sp.
Feeding on cantaloupe

closely monitored and use discontinued if food does not appear to be entirely consumed within a day or two. Powdered foods should never be used if there are any pest issues.

Liquid Nourishment: This can be brown sugar dissolved in water (3 parts water to 1 part brown sugar) or diluted real maple syrup. This is helpful primarily for very tiny, young millipeds in the first instars that are unable to travel far. A dropper full is squirted into the substrate every few inches so as not to be out of reach. Unlike the first three types of food, this does not rot or cause mite explosions. However, nematodes and springtails are very fond of it. Use is limited but good results are often visible.

Calcium Supplements: Cuttlebone and powdered calcium supplements have been commonly offered to captive millipeds. The milliped exoskeleton is primarily chitin, a long chain sugar polymer similar to cellulose. Calcium is usually not a major component of the exoskeleton, but it is a minor part of the matrix for the chitin that gives it rigidity. However, the primary rigidity is a result of the protein sclerotin and the small percent of calcium salts in the exoskeletons of any Juliformia would be reabsorbed prior to the molt. Unlike isopods which chew on chalk and cuttlebone in quantity to uncertain advantage, I have tested calcium blocks with various diplopods and have not seen any sign of ingestion. Cuttlebone and calcium blocks are primarily untouched by millipeds even when the cage is very overcrowded (at which point they chew on the plastic cage walls). Millipeds ingest substrate, so calcium-vitamin powders can be sprinkled on the dirt, every month or two, if feeding is desired. Powder additives, at times, seem to have a positive effect on growth and survival of immatures, but it may be the vitamins rather than the calcium. I have not used calcium additives (other than for testing purposes) over many decades, dozens of species, and tens of thousands of millipeds, so it cannot be a necessary additive by any measure. The tiny amount required is provided for by rotten wood or leaf mulch diets.

(Order Polydesmida)
Euryurus maculatus; Liberty Co., Florida
Captive-produced late instar
immatures surrounding dog food

Health Issues

"An ounce of prevention is worth a pound of cure," is key to maintaining the health of these animals in artificial habitats. Most damage can be prevented through careful attention to handling, enclosure design, and dietary intake with close observation of the animals' health and feeding habits. Minor health problems can heal following trauma caused by inattention and underfeeding, but even minor damage can require several molts and many years for full recovery. Major physical trauma (from drops or falls) either kills the animal or it survives, so prevention is the only treatment. There are issues that cannot be prevented because they are the result of old age or genetic defects.

One characteristic of millipeds making them great display animals and fantastic teaching tools, is that most are suitable for handling. However, handling must be responsible. Irreparable damage and death occur if an animal is dropped onto a hard surface. The exoskeleton cracks like an egg (chicken, not diplopod). Display animals should be held only above a surface not more than a foot away and preferably carpeted or cushioned. If the exoskeleton cracks and the animal survives the night, black scabs form on the cracks and the animal may live its full life. Even if there are no cracks, a dropped animal may still be found dead.

Though less common, improper habitat design can be the cause of damaging falls. Although many millipeds are excellent climbers, tall branches along with large rocks or bare bottom cages should be avoided. There is nothing positive to be gained from offering climbing areas, while cracks and damage are always a possibility. Rocks not only provide a hard surface to fall onto but also can smash an animal against the side or bottom of the cage when moved by the keeper. Escaped animals can fall to the floor as they leave a poo(r)ly designed terrarium.

Damage to the exoskeleton is most often caused by digging during necessary husbandry activities. It is important to be careful when digging through the substrate to locate specimens, rehouse, or to replace used-up or infested substrate. Small millipeds can be torn in half, while larger animals that have just completed a molt are readily smashed and ripped apart as well. Minor damage can heal, but freshly molted animals are so soft that trauma is far too massive for survival. Immature specimens molt regularly, so digging through a cage with hundreds of developing millipeds will almost certainly result in the death of a few individuals. The number of animals that will die if substrate is not replaced must be weighed against the damage it will cause. If there are no pests and there is enough room, food can be placed on top, and it may not be necessary to change out substrate for years. Shallower substrate helps to prevent digging damage, but surface molts and lack of space and food can be more detrimental.

Molts undergone by improperly fed specimens can result in massive body wall damage as the new exoskeleton forms improperly. Deformed segments (which look like the edges are eaten away) are commonly the result of underfeeding caused by poorly conceived substrate materials or pest infestations, rather than feeding dangerous materials (such as too much calcium or algae powders). I have seen this a few times in captive-reared specimens received from other breeders, so I cannot be sure what they were being fed. If the miliped survives, it can heal somewhat in subsequent molts, but rarely ever looks 100% normal again. Damage caused by molts that occur in collapsed cells look different—the bands appear misshapen rather than missing portions. With a good diet, this can heal in a single molt though scars may remain.

Overcrowding can lead to severe damage but is not necessarily related to the number of millipeds in the cage. Other terrarium cohabitants including isopods, cockroaches, and cage pests can cause overcrowding problems. In overcrowded conditions it can be difficult for specimens to form a molting cell. Lack of a cell can lead to deformities or cannibalism as the fresh exoskeleton is extremely soft. Chew marks may be seen on the edges of the segments as hungry tankmates are interested in the fresh exoskeleton and have no interest in the nearly indigestible exuvium. Extremely crowded habitats can be stocked with food constantly and still result in starvation of some inhabitants, while animals in underfed habitats may display the negative effects of overcrowding despite plenty of room. Shallow substrate, less than a few inches deep, can also result in overcrowding symptoms despite a large surface area.

Missing and damaged legs are commonly encountered problems with uncertain causes. 'Mycosis' or 'Foot rot' is often blamed, but most millipeds are designed to live in very damp surroundings—if the exoskeleton were terribly prone to fungal infections it seems the whole animal would rot away, not just the legs. Leg damage is found primarily on wild, adult specimens of species without post-ultimate molts. In captive-reared specimens this is sometimes observed on older adults that have been kept in cages overpopulated with isopods or, rarely, overcrowded with millipeds. However, even in these cases most specimens found in the same habitat have no damage and never on younger specimens. Various wild specimens have displayed this damage, so the cause cannot be known. (When imports were common, keepers often did not notice the damage for months and then thought the damage was new when observing the milliped groom itself.) It is possible other creatures in their natural habitats tried to eat them and grabbed a few legs. Leg damage can be repaired through molting, but it mostly occurs on older, mature specimens that are finished molting. Old insects often lose legs and feet as they age. Since millipeds have so many legs, even when more than half the legs are missing it has little effect on mobility.

As with damaged legs, damaged antennae grow back readily at the next molt, if there is one. A regrown antenna is often half the size of the original and is lighter colored or altogether lacks pigment. Animals missing one antenna seem to be unaffected, but if both are destroyed, they may refuse to eat, curl up, and wait for death.

Entomophagous fungi feed on the exoskeleton of living millipeds, resulting in death, usually within a few months. This type of mycosis has historically been restricted to just a few species (*Acladocricus* sp. 'Philippine giant blue,' *Archispirostreptus gigas*, and *Epibolus pulchripes*) and is lost within one or two generation of captive breeding. As with foot rot, immature specimens are not affected. It seems to be directly related to aging, but if the animals are kept on fresh substrate and are well fed, cases are less likely. Specimens can be treated with a fungicide if careful attention is paid to keeping it away from the spiracles (remember the ozopores are not the spiracles). However, there has been no evidence of treatment improving survival or longevity.

Pesticides pose some risk to millipeds because few, if any, millipeds can detect and avoid them. Usually this is not a problem because the substrate materials are decomposing, and modern pesticides break down with duration and moisture. Care should be taken when collecting materials. As with many problems, prevention is the only cure.

Congenital segment deformities (conjoined segments) are genetic and tend to have no effect on the health of the animals. These deformities do not heal through molting but are unlikely to cause molting difficulties past the second stadia. Similar deformities occur on various arthropods including beetles, cockroaches, and pseudoscorpions. It is best to avoid breeding these specimens.

Even when perfectly cared for, some millipeds will die. Much of the 'mysterious' die-off experienced can be attributed to cage pests and inadequately considered substrate materials, however different species can have vastly different survival rates under the same conditions. Die-off is a generally slow process that occurs over months and years (if they die all at once it is just death). Improper temperatures can lead to slow die-off if they are not terribly far out of the acceptable range. Many *Anadenobolus* die at a decent pace when kept a degree below 73° F (23° C), but otherwise seem relatively hardy. On the other hand, warm temperatures consistently above 80° F (27° C) increase the rate of die-off for most species.

(Order Julida) *Californiulus yosemitensis*

(Order Spirostreptida) *Spirostreptus* sp., adult; Tanzania

(Order Polydesmida) *Sigmoria trimaculata*

REPRODUCTION

The first step to encouraging reproduction is identifying and ensuring both genders are present. There should be at least one male for up to a dozen females, though a 50-50 split improves the odds of fertility. Determining the gender of millipeds is very easy with knowledge of the indicating characteristics.

Multiple mature males and females of a single species can be kept together without risk of the animals harming each other. Millipeds are seen as very docile, slow animals for good reason. However, when multiple mature males are in the same cage with receptive females, violent thrashing fights occur as rival males try to push other males off the females. If kept in plastic boxes, the fights can be quite noisy (especially with larger animals like *Archispirostreptus* and *Plethocrossus*). Fights usually occur only late in the evening and only during seasonal mating periods. Fortunately, fights do not usually result in damage (though gonopod damage is a possibility).

Adults and large immatures of most orders can be sexed with certainty in the middle to late stadia by the presence of gonopods or other male characteristics. However, defensive curling and defensive fluids can make it a challenge for smaller and less dimorphic species. For many species, sexual dimorphism makes sexing easy from a distance. One of the most distinctive is *Epibolus pulchripes*. The male has a glossy, jet-black body and oversized, fire-engine-red legs outfitted with large adhesive pads, while the female is dull, dark brown with small, dingy reddish legs. Many different species from different orders are easily sexed by size because the adult male is consistently smaller and shorter than the female. Gender determination for many round millipedes requires just a glance at the front of the body to look for the male's 'hood.' The male's body segments grow wider from the head back to the seventh segment, narrowing greatly at the eighth segment. The wide helmet followed by a narrowing of the body creates a hooded appearance.

In the round milliped orders (Juliformia), the male's legs located beneath the sixth or seventh segment back from the head are modified into gonopods. If there are no obvious color, size, or shape differences between male and female, the gender can be checked by looking at the legs beneath this segment. These modified legs (gonopods) are partly or totally withdrawn and often appear to be missing. If they can be seen at all, gonopods (secondary sexual organs) are highly modified and appear tiny, malformed, or like gelatinous sacs. The same

Gonopods
Note the missing legs on segment 7 and unusual form of legs on segments 5-6.

Adhesive tarsal pads are found on mature males of some species. This is a male *Epibolus pulchripes* seven days after the ultimate molt. It takes about thirty days for the legs and face to darken to dark red after the ultimate molt in this species.

(Order Spirobolida)
Epibolus pulchripes
3rd generation, juvenile color phase
Immatures between and one to two inches have red telsons.

(Order Glomerida) *Glomeris klugii*, male claspers

(Order Polydesmida) *Antheromorpha uncinata*, mate guarding
Males of this species can guard the female for up to eight weeks continuously.

(Order Polydesmida)
I collected some tiny, early instar *Euryurus* and *Oxidus gracilis* from under a log in Muskingum County, Ohio, and raised them up. Recently I saw two *Oxidus* that looked like they were eating a dead *Euryurus*, but on closer inspection the two male *Oxidus* were fighting for mating privileges.

(Order Polydesmida)
Two male *Oxidus* fighting for mating privileges with a female *Euryurus*.

(Order Spirobolida)
Chicobolus males are identifiable by the wider segments six and seven. Two predatory mites, *Stratiolaelaps scimitus*, are on the substrate, on both sides of its face.

(Order Spirostreptida)
Tamil Nadu, India
Amar Bharathy (CC0)

(Order Polydesmida)
Orthomorpha subkarschi
Courtship

(Order Polydesmida)
Paradoxosomatid millipeds
Sri Lanka
Maximilian Imran Faleel (CC BY 2.0)

pairs of legs are not missing and look normal on the female. The missing legs are often very noticeable from the side, though they leave a small gap which can be more cryptic in spirostreptidans. Gelatinous sacs when extruded are visible from the side on males at segment seven and females at segment two. The presence of gonopods can be used to determine gender even when giant round millipeds are less than 1/3 full grown (though gonopods must be fully mature for use in taxonomic identification). Though sexing immatures is technically very easy, it can be dangerous because small animals are more delicate and often in pre-molt or post-molt. The exoskeleton is soft after every molt and forcibly unrolling the front of the body can cause severe damage and death.

Giant round millipeds wrap slightly around each other during mating and the male extends the normally hidden gonopods as he embraces the front of the female in a secure 'hug.' His gonopods on the seventh segment attach to her vulvae on the second segment which are also everted for mating. It is easy to tell which is the male and female by their position during mating. The embrace lasts from fifteen minutes to a few hours, but can recur every few days or weeks. In a few species, as the male holds the female he rhythmically pulses all the legs along his body during copulation.

Flat millipeds are sexed in the same way as the juliform orders. Males also have gonopods located on the seventh segment back from the head, but only the front pair is modified. These show up as odd little appendages in the middle of an aperture. It is much easier to see than the normally hidden spirostreptidan gonopods. The female's vulvae are located on the second segment behind the base of the legs. During mating, polydesmidans can assume a unique 'pinwheel' configuration or wrap together at the front like round millipeds.

Males of the Pentazonia orders (pill millipeds) have modified legs at the end of the body. Females have more legs, all normal, while males have fewer legs with the terminal pairs modified into claws. The female has nothing like those. The 'claws' are used during mating and in some species may make noise. Unfortunately, these appendages are only visible if the animal is open—at which point it is either dying or dead. (Healthy animals stay curled up many minutes to many hours after being moved or

(Order Spirobolida) *Apeuthes* courtship

(Order Platydesmida) *Petaserpes* courtship

(Order Sphaerotheriida)
Arthrosphaera sp.
Female ventral posterior

(Order Sphaerotheriida)
Arthrosphaera sp.
Male ventral posterior

(Order Polydesmida)
Euryurus sp.
Muskingham County, Ohio
Mate guarding

(Order Spirobolida)
Wild *Aphistogoniulus* mating

handled.) Fortunately, most species exhibit an obvious external difference. The large, flattened terminal segment (anal shield) of the female is smooth and unmarked. The male *Arthrosphaera brandtii* pill has a large flattened 'O' shaped marking in the middle of the anal shield. The Madagascan *Zoosphaerium* male has a sculptured 'V' pattern. These markings make gender determination very easy, though sometimes the markings on the male 'Tanzanian' are barely visible. Species of *Zephronia* from Thailand also exhibit gender differences in the shape or color pattern on the anal shield. Sexual dimorphism of this segment may be a standard feature of the sphaerotheriidans.

The Tanzanian and Madagascan pill millipeds that had been common imports were extremely gender biased. Approximately one out of every ten wild-caught specimen was a male and finding zero males among dozens of animals was not a rare event.

Giant pill millipeds have a unique mating posture due to the design of the body. Both mates of course must be uncurled, but the male curls up partly under the female, backwards and upside-down. If his reproductive appendages were near the front of his body like most millipeds he would be screwed (or more literally not, *not* screwed since male diplopods employ accessory organs versus genitalia).

Following courtship, the females of even the most unusual type of diplopod store the spermatophore inside spermathecae. The sperm will be used to fertilize the eggs when they are laid. Females can store sperm for at least a month, though reproductive experiences with albino *Narceus* discussed in the species list suggests a year is possible.

(Order Polydesmida) *Cherokia georgiana latassa*
Dekalb Co., Alabama

(Order Polydesmida) Bioluminescent *Motyxia sequoiae*
California

(Order Glomerida) *Onomeris* sp., mating
Alabama

(Order Spirobolida)
Centrobolus sp.
Mozambique "Crimson Blackleg"

(Order Spirobolida)
Pelmatojulus excisus
Note the gonopods on the seventh segment of the male are inserted into vulvae on the second segment of the female. There are only two millipeds in this picture.

(Order Spirostreptida)
Spiropoeus fischeri
Third generation, mating

(Order Sphaerotheriida)
Arthrosphaera brandti
Mating

(Order Polydesmida)
Orthomorpha subkarschi
Mating

(Order Spirobolida)
Trigoniulus macropygus
Fifth generation, mating

Parthenogenesis

Some members of the Orders Julida, Polydesmida, Pselaphognatha, and Spirostreptida are known to reproduce through parthenogenesis. Sexing and mating are unnecessary. Still, this form of reproduction is not widespread and is extremely rare for the larger millipeds. In captive breeding situations only a single large spirostreptidan has laid eggs without the recent presence of a male, the 'earthworm milliped' *Macrolenostreptus brachycerus.* Males and females are often found in equal numbers for wild-caught and cultured *M. brachycerus,* so this is facultative rather than the primary method of reproduction. The high female to male ratio for some sphaerotheriidans and the following *Narceus annularis* anecdote suggests parthenogenesis could be a secondary reproductive method for some species.

Normally, wild and captive specimens of the large captive polydesmidans, spirobolidans, and spirostreptidans are approximately 50% male and 50% female (gender is random so the larger the group of animals, the closer the ratio is to 1:1). Gender imbalance of wild imports might be the result of parthenogenesis in sphaerotheriidans. It is common for giant pill millipeds in the genera *Arthrosphaera* and *Zoosphaerium* to come in large groups of animals that are 100% female—never more than 20% male. These male to female ratios are based on hundreds of wild-caught imports. It is possible there is a natural gender bias or seasonal activity leading to disproportionate numbers, or males may just be short lived. From experience with a small group of *Zephronia siamensis,* there was one male in four, which died after two weeks.

In one of my first *Narceus* breeding projects in 1994, a wild female *N. annularis* found in Brunswick, Ohio, mated with a wild male *N. americanus* collected in St. George, South Carolina. At the time I believed both were *N. americanus.* The female *N. annularis* was mature and was provided egg-laying stimulus, but it did not lay eggs over a five-month period. The habitat and substrate were unchanged when the South Carolina adult male was introduced. The two mated repeatedly and the female immediately began to form egg capsules. It took a few years of growth before the young could be sexed and by that time most had been traded off. After four years, all thirty-one remaining adult offspring were female. This would be realistically impossible if the gender split were 50/50. (Consider the chances of flipping a normal quarter and getting 'heads' 31 times in a row.) It is possible the female used this male's sperm to fertilize her eggs, but the sperm were unacceptable (and ova developed through parthenogenesis) or only the female gametes were able to successfully fertilize. Later laboratory reproduction events and a wild-population count of thirty *N. annularis* from northern Ohio resulted in an approximate 50/50 split on gender. Males from later broods did not experience a higher die-off or grow at a different rate that might have otherwise explained the gender bias associated with the interspecies breeding. *Narceus annularis* coloration even within a population is variable—primarily the shade and intensity of the pink/red—but the 31 females were nearly identical. Consistent coloration is anecdotal since mantids and phasmids produced through parthenogenesis can develop different colors when raised in the same artificial habitat. One point of interest is the total die-off of eggs was not noticeably different between this and other broods. (If half the eggs were unsuccessfully fertilized by male gametes the die-off ought to have been significantly higher. Parthenogenesis, when not the norm, can also result in lowered hatch rate.) Total or partial unsuitability of the gametes was most likely the result of the two being different species. Female *N. annularis* do not lay eggs without recent male contact, so the *N. americanus* seems to have provided the stimulus for egg laying but not the fertility.

Egg Laying

Once the adults have been sexed, the next step is to provide an environment for egg laying. With both sexes kept in the same habitat, fertility is virtually assured since they present none of the mating challenges associated with predatory or short-lived arthropods. Mating is more often observed when adults are healthy and within acceptable egg-laying environments. Enclosure materials and habitat parameters (humidity, plant materials, temperature) may need to be adjusted if no mating is observed. However, observed mating does not ensure egg laying.

EGG LAYING MEDIA QUALITIES

• Substrate depth should range from 3-6″ (7.6-15.2 cm). Shallow substrate can result in adults damaging the eggs and smashing protective capsules. Limited depth can also reduce or prevent egg laying. Small species, as well as polydesmidans and glomeridans, only need an inch or two (2.5-5.2 cm) to maintain a humid environment. Surface area is also important—a container should be at minimum twice as long as the longest adult in both length and width. It can almost never be too large.

• Moisture level of the substrate should be damp, but not oversaturated, or it becomes anaerobic. Eggs have zero tolerance for dryness and females cannot form capsules or excavate compartments in dry substrate.

• Finely ground materials should make up 50% or more of the substrate. By itself, coarse substrate such as wood chunks or large leaf pieces cannot contain eggs. Finely-ground rotten wood, shredded decayed leaves, or composts mixed with expanded oak pellets can work well. Potting soil and peat have the right consistency, but present little nutrition and are likely to starve small immatures.

• Approximately one part ground wood and one part leaf particles are mixed with two parts of the base. Although not a necessity for the initial egg laying, this is important to prevent starvation as food must be easily accessible for very tiny larvae. Base substrate should be primarily rotten wood for early larvae of *Centrobolus* and the Puerto Rican *Anadenobolus*. Polydesmidans should be provided strictly rotten wood or quality wood flake soil for best results.

• An egg-laying stimulus ranges from helpful to necessary, depending on the species. A layer of dead hardwood leaves and chunks of decaying wood are needed for most juliforms: 50-100% rotten wood for spirobolidans and 75%-100% leaf litter for spirostreptidans. A 1-2″ (2.5-5.2 cm) layer of leaves or shredded rotten wood are placed on top of the base. Decayed leaves from beech and elm are excellent, but the leaves of almost any flowering plant can work. A faint layer of white mold attacks newly wetted leaves and may be more of a stimulus than the leaves themselves. Leaves that have been continuously kept wet for months may not work. Rotten wood should be soft enough to be easily chewed by the milliped. If the adults are mating but there are no signs of eggs, try adding leaves or wood at a different level of decay or from a different area. For pentazonidans and polydesmidans, the ground algae and detritus below small-leaved mosses can be useful impetus.

The presence of a male (recent mating with) may be an additional, important egg-laying stimulus for at least the spirobolidans and spirostreptidans. Nevertheless, it is a difficult-to-prove theory. I have attempted, on numerous occasions over the years, to collect eggs from various females by providing a good egg-laying habitat at the right time of year when no males were available. With the exception of the facultatively parthenogenetic spirostreptid *Macrolenostreptus* and the polydesmidan *Pachydesmus,* no others have even considered laying a single egg if they had not mated in the last two months.

Most female spirobolidans and spirostreptidans produce one, rarely two, batches a year (commonly 200-500 ova). If more than a year goes by without eggs, the animals are either very unhealthy, immature, or the proper stimulus has not been provided. Laying usually takes place one egg at a time, over months, though at least two species can deposit hundreds of eggs in one big mass on the same day.

Spirobolids with annual molts tend to lay eggs within two months of the molt and may not lay eggs after that time before the next molt. *Chicobolus spinigerus*, *Tylobolus uncigerus*, *Narceus americanus*, and *Narceus gordanus* tend to lay eggs only after the annual molt, while *Narceus annularis* does not coincide egg laying with molts.

Nearly every species with a captive history in the sprirobolidan and spirostreptidan orders has been bred in captivity. Only recently has there been some success with members of the Order Polydesmida as a result of better understanding of the egg-laying requirements. In nature giant pill millipeds form chambers carved in the soil, under a rock or log, or inside rotting wood. Chambers should be the size of a curled-up female and may contain a few hundred eggs. Unfortunately, keeping giant pill millipeds alive for long periods is difficult, so reproduction is an afterthought.

There are a number of ways females protect their eggs, which can vary as much by species and genus as by order. Egg guarding is rare, but guarding through hatching and the second stadia is documented for females of one spirostreptidan species (McMonigle, 2005) and for males of multiple species in the Order Platydesmida (Hoffman, 1982). Egg guarding is unusual for polydesmidans, but *Platydesmus complanatus* and *Pseudopolydesmus canadensis* females create a frass cell, lay a mass of eggs, and wrap their bodies around it before dying (Stanek, 1969). The peculiar polyxenids coat the eggs with the setae from their tail (Forster & Forster, 1980). Many females from various orders construct hollow dirt spheres and glue a single egg to the inner wall. This can be done from the inside or outside, compacted by the anal valves, and the construction materials can vary. The large round millipeds primarily use their mouths and front legs, glomeridans the rear, and polydesmidans use their rear ends to build everything (old cages look like crazy cities of frass). These shelters protect the egg and protonymph from the environment, but it seems the better the shelter the more likely the egg cannot survive without it.

Each species may have a slightly different take on egg laying and protection, but the methods follow a limited set of possibilities. Variations observed in captive specimens are described below.

SPIROBOLIDAN EGG LAYING

1. The most common method is the familiar milliped egg capsule. Well-rotted wood is the usual stimulus for capsule formation, but capsules may be formed from most particulate substrates including decayed leaves and compost. The female creates hundreds of thin-walled capsules, each with a large central chamber and a single egg attached inside. During a four-to-six-week period she constructs one capsule at a time and can always be seen at work—the brood will hatch out over about the same period of time. She uses her mouth and the legs on her first ten segments to form the capsule from chewed wood and substrate she collects. The legs hold the capsule walls in place while she compresses and smooths out the inner surface using her gnathal lobes. She glues the egg to an inside wall before finishing construction of the capsule. After completion she can use the anal valves to compress the capsule and may also adjust the moisture.

Normally the egg is extremely small, so the hollow center of the capsule is many times the size of the egg. There are exceptions; the monstrous eggs of *Epibolus pulchripes*, *Pelmatojulus* spp., and *Atopochetus dollfusii* fill up a good portion of the empty space. In some species the egg capsule is approximately the same size and shape of an adult's frass pellet (as seen with *Acladocricus*, *Apeuthes*, *Centrobolus*, *Chicobolus*, *Epibolus*, *Narceus*, *Trigoniulus corallinus*, and *Tylobolus*) while with *Pelmatojulus excisus* and *Aphistogoniulus* the egg capsules are much larger than frass. In *Trigoniulus macropygus* the capsules are strangely dimpled and attached to a conical frass pellet. *Acladocricus* sp. 'Philippine giant blue' will create some egg capsules, but they are fragile and most of the eggs encountered are bare.

The resemblance to frass is commonly cited as a form of camouflage and piles of egg capsules readily disappear within mounds of frass. However, in the wild, capsules are found inside hollow logs and under bark where they are not visible. Exposure would kill the eggs and subterranean creatures are not known for camouflage. The visual resemblance is partly

(Order Spirobolida) *Aphistogoniulus* sp.
Sealed egg capsule, opened capsule, and egg
Egg in opened capsule is adhered to the capsule wall. Note the capsule is formed from wood fibers whereas frass often looks like brown toothpaste.

(Order Spirobolida)
Apeuthes sp.
This egg seems to be floating since it is glued to the wall of the capsule.

(Order Spirobolida)
Trigoniulus macropygus
The unique conical fecal pellet is not eaten by the larva and can be removed easily without damaging the egg capsule (and the frass cone occasionally falls off). The unusual, large pits on the surface of the capsule (none on the frass pellet) are obscured by flash photography.

(Order Spirobolida)
Narceus americanus albino cultivar
Fifth generation
Egg capsule with side removed

(Order Spirobolida)
Acladocricus sp.
1st instar larva exiting capsule
Egg capsules are abnormally difficult to differentiate from frass unless one is accidentally smashed. Capsules have a thin wall and are mostly hollow which allows for the great increase in size from egg to protonymph and then 1st instar.

(Order Spirobolida)
Acladocricus sp.
Early protonymphs and a 1st instar

(Order Spirobolida)
Anadenobolus arboreus
St. John, U.S. Virgin Islands
Eggs

(Order Spirobolida)
Aphistogoniulus sp. at six months

(Order Spirobolida)
Aphistogoniulus sp.
1st instars and egg capsules
Spirobolid 1st instars have one ozopore and six legs on each side. This is the stage where larvae begin feeding. Frass is visible inside the abandoned egg capsule.

(Order Spirobolida)
Chicobolus spinigerus
Early stadia

(Order Spirobolida)
Six egg capsules and a bunch of frass
Capsules here are obvious because of the texture and color. (Color is often the same unless somewhat dry.) One is opened for illustration.

(Order Spirobolida)
Pelmatojulus excisus
~ 1" (2.5 cm) and one year old

(Order Spirobolida)
Trigoniulus macropygus
Immatures, fifth generation (2009)

(Order Spirostreptida)
Spiropoeus sp.
Immatures at 30 months
Adults were acquired under the label 'Archispirostreptus' in 2001.

(Order Spirobolida) *Acladocricus* sp.
Early stage protonymph

(Order Spirobolida) *Acladocricus* sp.
Older protonymph, ready to molt

(Order Spirobolida) *Pelmatojulus* excisus
1st instars are extremely fat because the eggs are very large and the space inside the capsule will not allow for flattened development.

coincidence, but also is related to the way the capsules are formed since the anal valves are often used in creating or solidifying and adjusting the moisture of the capsules. On close inspection, even capsules the same size and shape of frass are visibly different. Egg capsules have a uniformly fibrous surface and are nearly perfect spheres. Frass has a smooth surface and is an irregular, oblong, or deformed ball. Unlike polydesmidan cells, these are formed from undigested substrate, not frass, and are not cemented to surfaces. Eggs laid in this way are usually extremely sensitive and die if removed from the capsule. They may survive if only a hole is made in the capsule, but damage to capsules usually results in death.

2. Eggs are laid individually in the substrate with no capsule. Poor conditions can cause sporadic laying, but consistent care results in egg laying and hatching stretching out over years. This method is found in the spirobolid family Rhinocricidae and has been observed for most of the *Anadenobolus* that have reproduced in captivity (see species list). Females drop eggs in practically any fine, damp substrate. One egg is laid at a time, from a few a day to one every few days. A healthy, mature female can keep this rate of laying up indefinitely, when quality food and a mate are provided. Eggs are tiny, translucent white and appear fragile, but survival is good without egg capsules. Wood and leaf components are not required to induce egg laying, nevertheless they are important to keep the young alive and growing.

SPIROSTREPTIDAN EGG LAYING

1. Eggs are commonly deposited naked in the substrate, sometimes in a small airspace.

(Order Spirostreptida)
Archispirostreptus gigas
Early stage protonymphs within egg shells (F3, 2016)
These were gathered and placed on top of the substrate for observation.

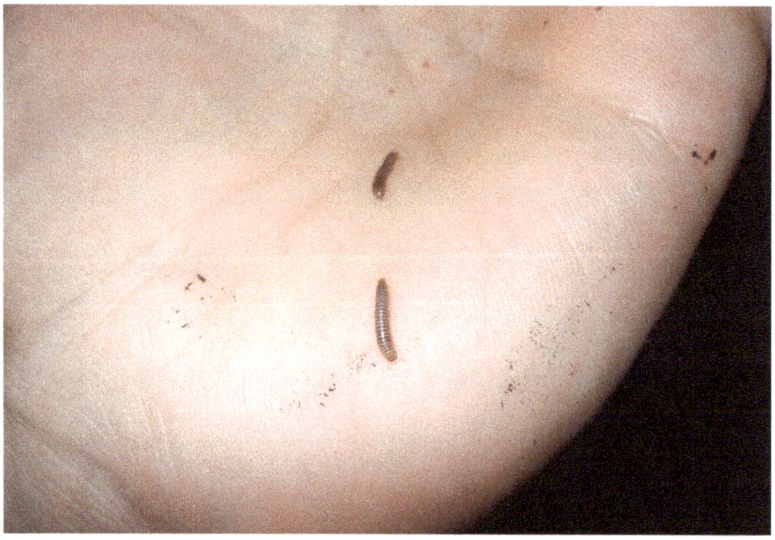

(Order Spirostreptida)
Archispirostreptus gigas
2nd instars
Offspring from captive-born parents are rare. Letting a wild-caught female drop eggs is a far cry from rearing and conditioning adults.

(Order Spirostreptida)
Spiropoeus fischeri
Eggs seen through glass from beneath terrarium

(Order Spirostreptida)
Dendrostreptus macracanthus
Eggs from one female and 1st instars from another (F1, 2004)

(Order Spirostreptida)
Dendrostreptus macracanthus
At nine months (2012)

(Order Spirostreptida)
Dendrostreptus macracanthus
Second generation female excavated from brooding cell (egg mass accidentally broken up and partly buried)

The laying and hatching periods are usually stretched out over a month or two. The ovum is normally large (1 mm or more across), round, and yellow to brown. The stimulus for laying is commonly leaf litter. Millipeds that employ this method include *Archispirostreptus, Macrolenostreptus, Spirostreptus gregorius, Tropostreptus hamatus,* and an unidentified African crested species with white legs. The egg membrane is thick and very noticeable when shed by the protonymph—it may persist in the substrate for months or years. These eggs are very sturdy because they do not require protective egg capsules to gestate successfully. Development times and stadia can be harmlessly observed. Depending on species, a small percent of eggs may be found in difficult-to-define, flimsy, attached constructions, but nothing consistent or observably separate from the substrate.

2. Eggs may be deposited in a large pile buried in the substrate. Eggs are laid and usually hatch the same day. *Spiropoeus fischeri* deposit their eggs a few inches underground, usually against the hard bottom of the enclosure. *Dendrostreptus macracanthus* likewise forms a huge mass of eggs but these are watched over by the female. She forms a large cell in the substrate similar to a molting cell, but wraps her legs around the eggs and stays with them until they have hatched into protonymphs and molted to first instar. If disturbed, she will not regather the eggs or developing stadia. Eggs of both these species have thick shells and develop successfully without a mother's attention.

3. Female *Plethocrossus* sp. use the mouth and anterior legs to construct a large, flattened, triangular, hollow container for each egg. Unlike common spirobolid capsules, these would be far too large to be adjusted or compressed by the anal valves. Not only are they unable to fit between the valves, but there are often objects sticking out the side. These capsules should not be opened because exposed eggs die within days.

POLYDESMIDAN EGG LAYING

1. Eggs can be laid in individual cells throughout the habitat somewhat like spirobolid method 1, however these structures are flattened and pasted inside clumps of frass attached to rotten wood or mounds of frass. Capsules usually collapse if separated from the substrate, since they are pasted into the substrate. The cell is two or three times the size of the egg. Multiple cells can be constructed each day, but the process takes a month or two and the young hatch out sporadically over a similar length of time. The stimulus is well-rotted wood. *Euryurus leachii* and *Euryurus maculatus* employ this method. Exposed eggs can still develop if covered lightly with frass.

2. Small paradoxosomatids including *Helicothomorpha* and *Oxidus gracilis* lay multiple small clusters of eggs in damp substrate. Decaying leaves, wood, and certain algae can stimulate laying.

3. A globular mass of hundreds of eggs is deposited inside a large cell that is created from many hundreds of frass droplets. Construction is similar to the molting cell and the cavern is about fifty percent larger than the mass of eggs it contains. Considering the volume of frass needed to construct the cell, it likely takes the female a few weeks to build it. When this method is employed the young, which can number in the hundreds,

(Order Spirostreptida)
Plethocrossus sp.
Tanzania import (2002)
Three frass pellets (left) compared to three egg capsules (right, one opened)

(Order Polydesmida)
Euryurus leachii
Northeastern Ohio
Immatures

(Order Polydesmida)
Pachydesmus crassicutis crassicutis
50+ hatchlings
This depicts 25% of the young that emerged from one egg mass on the same day. They are fertilized, laid in mass, and emerge together.

emerge the same day. Eggs are much smaller than in those species laying eggs in individual cells, because the female here develops all the eggs at once. *Pachydesmus crassicutis* lays eggs in this manner. It is many times the size of *Euryurus,* but the eggs and early instars of *Pachydesmus* are much smaller (McMonigle 2010). One female can lay more than one of these batches over time.

4. Large cells are formed in the substrate made from frass (similar to 3), but the female seals herself inside with them. If disturbed before laying eggs, the female tends to die without ever ovipositing. It is unlikely a female could ever lay more than one batch of eggs.

GLOMERIDAN EGG LAYING

Glomeris and *Rhopalomeris* lay eggs one at a time within hollow capsules, like spirobolid method 1, but they are less easily separated from the substrate. Egg capsules are produced over a few months once or twice a year, or continuously under good conditions in captivity. Exposed eggs usually die. The impetus for laying is the detritus and cyanobacterium that builds up under small-leafed mosses such as *Leucobryum*. Capsules are laid in the surrounding substrate.

(Order Glomerida)
Glomeris pustulata
Egg

SPECIES LIST

Most listed species in the Orders Spirostreptida and Spirobolida have been and continue to be successfully bred and reared. Order Polydesmida has seen limited species-specific success, while the Order Sphaerotheriida is likely to never be reared with repeatability. Order Glomerida has been successfully reared through generations only recently. Order Julida has been reared in huge quantities as habitat pests and are often mistaken for baby spirostreptidans. The very rare large julidans, Family Paeromopodidae from California, made it on the rare vendor list starting in 2021 and appear to be much more challenging. Order Platydesmida can be maintained and bred under laboratory conditions.

Common names listed here are the names with regular use in the pet trade. Unfortunately, many collected specimens have not been identified and do not have verifiable scientific names, though identification to the genus is usually possible. The date and descriptor name are not part of a scientific name and do not imply accuracy. Nearly all descriptor names are in parentheses because the genus was changed after the original description.

Characteristics, specifically coloration, but also size and gonopod structure, can be highly variable. Size listed is the maximum verifiable or typical adult size using vernier calipers for measurement. Size estimates by those who choose not to employ a measuring device average twenty to fifty percent inflated lengths over those listed here.

Time required to reach adulthood assumes adequate food, consistent moisture, and a temperature of 70°-76° F (21°-24° C). Poor but sublethal conditions may greatly lengthen the reported developmental times.

Order Spirostreptida

These are often the longest and bulkiest millipeds in tropical and subtropical arid lands. Many have close to three hundred legs and seventy body segments. The legs on the male's seventh body segment are highly modified for mating. Though some adults are contrastingly colored, even the brightest tend to be in earthy tones.

African Giant Black

Archispirostreptus gigas (Peters, 1855)

The African giant black (AGB) is the largest milliped in the world. The verified record specimen at 15.2″ (38.6 cm) is an *Archispirostreptus*, probably *A. gigas*. Most do not grow so large: males are usually 7-8″ (17.8-20.3 cm), while females are commonly 8-9″ (20.3-22.9 cm). To the novice, even a seven-inch-long AGB appears to be a foot long. The body is as thick as a cigar. The sleek, powerful appearance impresses even the non-invert fan. With other invertebrates, the largest species is generally the most expensive and difficult to acquire, and this has become the case for AGBs. Since the importation ban, tubs overflowing with wild-caught adults or captive-bred juveniles for a few dollars are no longer a common sight in North American pet shops or swaps. *Archispirostreptus* are still very common in European swaps as imported adults from Kenya to South Africa. Some imports are likely from related species in the genus, or even unrelated genera.

The powerful, gentle feel of an AGB's grip as it walks across your arm is strangely compelling. They are not venomous and cannot inflict damage from bites, pinches, or scratches. Still, they are not completely defenseless. Frightened AGBs coil tightly or discharge toxifabricants from the ozopores and eject watery frass from the paraprocts. AGBs are escape artists because they are big and strong. If the lid is not locked tightly on the plastic box, or the glass lid on the aquarium is left ajar, the exit is likely to be quickly discovered.

Females are picky and will not drop eggs just anywhere. The substrate must be damp for egg laying—it can be as shallow as 2″ (5 cm), but 4-6″ (10-15 cm) depth works best. Potting soil, coconut fiber, rotten wood, and compost are all acceptable substrates. They refuse to lay eggs, or lay just a couple, if there is no stimulus. A 1-2″ (2.5-5.1 cm) layer of old leaves added to damp substrate can be all that is needed to encourage a healthy adult female to begin egg laying. Decaying leaves (a mix including oak and beech) should be chosen that have been on the ground a few months or more. Very dark brown to black leaves is an acceptable food, but a poor stimulus.

Females deposit 0.125″ (3 mm) yellowish-white eggs a few inches deep in substrate. They do not construct sturdy egg capsules or lay in compact nests. As common for spirostreptidans, eggs are laid bare in the substrate and have a thick, durable shell. Still, eggs can be smashed and should be removed from the adults' enclosure for the best survival rate. Eggs are tough, but early larval stages and molting periods are delicate. They are laid one at a time, but over a few months a female can produce hundreds. Eggs are produced once a year, usually September-November for wild-caught, but captive-reared females have laid eggs in January. The fertility of eggs from captive-reared specimens is sometimes low.

The tough eggshell splits to reveal a protonymph after about a month. Although the protonymph is bright white, the yellow-white egg often darkens to red-brown during incubation. The protonymph resembles a tiny croissant and is motionless, legless, and without visible ocelli. It is more vulnerable than the egg and easily trampled by adults. (The egg shell is very thick and can be dug up many months after hatching.) After a month the protonymph hatches into a tiny white milliped with legs only on three of the front body segments. Though it has very few legs, it burrows nimbly into the substrate to feed. In three to four weeks it molts and gains legs on all but the last few segments. A quarter of the eggs make it past the first molt, but females lay countless numbers, so the survivor count is generally still in the hundreds. Die-off after this stage is minimal. Adults do well at warmer

Archispirostreptus gigas (left)
Spiropoeus fischeri (right)
Mature females

Archispirostreptus gigas
Kruger National Park, South Africa
Bernard Dupont (CC BY-SA 2.0)

Cambala annulata
Adult with three-month immature

temperatures, 77-82° F (25-28° C), but immatures should be kept cooler.

Six months following hatching the milliped is pale-gray and only 0.75" long. It grows a little darker at each molt and adds a few segments. After six or seven molts the number of legs and body segments approach mature quantities. At two years it should be dark brown and 3-4" (7.6-10.2 cm). AGBs average five years (ranging 3-7 years) to reach adulthood and can live a decade or more if properly cared for. Captive-reared specimens resemble wild adults, but they are often a little lighter in color, possess minor segment blemishes, and do not have commensal mites. Wild adult AGBs have been kept for seven years without a single molt. There are a few reports of 'possible' adult molts, but these have not been for specimens that have reproduced. If an adult seems sluggish, like it is about to molt, it is more likely to die.

Amethyst Crested Milliped
Cambala annulata (Say, 1821)
This heavily sculptured North American species ranges from Indiana to western Pennsylvania, south along the inner coastal plain to northern Florida. Adults are violet gray and grow to 1.5-2.25" (4-5.7 cm) in length. Six other species in the genus from North America are: *C. hubrichti, C. minor, C. ochra, C. speobia, C. texana,* and *C. washingtonensis.* Of these, *C. minor* is found in many of the same areas. It looks similar but is notably smaller and paler and more likely to be found in groups.

Cambala annulata

Cambala spp. are sometimes confused with *Abacion* spp. from the Order Callipodida. The exoskeletal sculpturing of callipodid(a) lends more of a velvety appearance, but sometimes they are given the name 'ridged' or 'crested.'

Cambala annulata is host to a specific micro fungus, *Troglomyces twitteri* Enghoff and Reboleira 2020, first identified through a picture posted on social media by Derek Hennen in 2018. The specimen was a male from Adams County, Ohio, photographed in 2014. Although the image was greatly magnified, it seems amazing someone would notice the two tiny structures that look like particles of dust. That person must have been studying microscopic fungi associated with arthropods. I often notice commensal mites on millipedes since they are an interest of mine, but mites have a more obvious shape and the micro fungi are so much smaller.

The specimens depicted here are from Hocking County, Ohio. Wild *C. annulata* are commonly found individually, as they are generally solitary animals. However, up to a dozen adults can be found clustered together under logs during dry periods.

Cambala are active and heavy feeders on almost any fruit and some vegetables, but only very young specimens normally consume substrate. Like many spirostreptidans, they seldom nibble on rotten wood or leaves in the enclosure, while melons and dog food are eaten with some zeal. *Cambala* are among the rare millipeds that eat raw carrots.

Adults collected in mid-October laid a dozen eggs in an artificial habitat. Pairs continued to mate intermittently through December, then stopped. The eggs were laid intermittently in shallow substrate. The small number of eggs may have been related to the shallow substrate (less than an inch, or 2 cm) or lateness in the season. They hatched and developed through second stadia in four weeks.

Choctella cumminsi Chamberlain, 1918

This is a short, 1.5″ (<4 cm), but thick-bodied spirostreptidan found only in central Tennessee and northern Alabama in the United States. The segments are dark brown with a wide, patterned 'stripe' along the dorsal surface of the body. The stripe is an opalescent patchwork of white to reddish-brown spots. Adults spend a few months on the surface eating, mostly melon and rotten wood in small quantities, and then form individual pupal cells at about the same time. Synchronous pupal cell formation is a common spirostreptidan behavior, but the molting period of *C. cumminsi* is unusually slow. The inactive period for late stadia molts is about six months from pre-molt to post-molt. This is the only spirostreptidan I am aware of that appears to go through regular post-ultimate molts.

Glossy Black Pinkleg

Dendrostreptus macracanthus (Attems, 1914)

This lengthy species from Mozambique to northeastern Tanzania is one of the more brightly-colored spirostreptidans and one of the largest kept in captivity. The body segments from head to telson are glossy black like patent leather. The legs are long in relation to the body width and are a delicate pastel pink with pale pink banding. Males are seldom over 6″ (<15.2 cm). Females grow to 8″ (20 cm), making them the longest of any commonly kept species other than the familiar African giant black. Nevertheless, the body is thin, so a couple of shorter, thicker species are more massive.

Males are always noticeably smaller by about 30% and are easily sexed from above by the presence of an obvious hood. They battle over females, but mating is not seen often and generally occurs in the late evening. During mating the male rhythmically squeezes and releases all of his legs along his entire body, about once every two seconds. The leg-pulsing behavior does not serve an obvious function but has been observed in some other species including *Spirostreptus gregorius* and *Epibolus pulchripes* (a spirobolidan).

Adult *D. macracanthus* are primarily arboreal in the wild and almost never encountered at ground level (Hoffman, pers. comm. 2005). The enlarged legs and bright colors can be easily attributed to life in the trees. They also have two behavioral traits likely to coincide with an arboreal nature. First, specimens readily secrete more defensive fluid than other species, which would be valuable for a constantly exposed animal. Second, all the eggs are laid in a single evening, in contrast to the weeks to months many other spirostreptidans spend ovipositing individual eggs. Despite arboreal habits, captive-caging elements used for terrestrial species work well and immatures build underground chambers for molting.

At most, one time a year, the female builds a loose, shallow cavern and deposits up to a few hundred yellowish, 1.5 mm

Choctella cumminsi
Feeding on bark

Choctella cumminsi
Molting

Choctella cumminsi
Lincoln Co., TN
Jay Brasher

Dendrostreptus macracanthus
Wild-caught adults

Dendrostreptus macracanthus
Subadult, third generation

Dendrostreptus macracanthus
Immatures, third generation

eggs in a large, slightly sticky mass. (They appear stuck together but the mass collapses if touched.) Females housed in the same enclosure produce batches of eggs within a month of each other. A mix of primarily oak and beech leaves, which had been on the ground at least half a year, was used to induce egg production one year. The following year, maple leaves that had been collected in fall and stored damp in a trash bag for two years were used. The number of eggs laid the second time were minimal, but the low numbers may have been caused by age or possibly impacted by the number of immatures still present from the previous year. Second generation adults produced 200-400 eggs per brood. There were only seven fourth-generation hatchlings due to low egg production and low hatch rate.

Larvae look very glossy even when small and pale. Young are hardy and do not seem to experience the usual noticeable die-off resulting from insufficient feeding. They grow at a good pace and some reach two inches within nine months of the eggs being laid. However, adulthood will only be reached after four to five years. Once the immatures grow to a few inches in length, all individuals in a cage will molt together even if they are different sizes. After growing to adulthood, they live another three to four years. One captive stock (pictured) was bred in North America with some regularity from 2000 to 2023, but there may only be a single male left.

Tanzanian Earthworm
Macrolenostreptus brachycerus (Gerstaeker, 1873)
This was among the suite of Tanzanian import species in the 1990s. The specimens depicted here are descended from adults imported from Tanzania in 2002 (through International Dragon Breeders) and the colony is still going strong in 2023. Adults are moderately large and grow to 3.5″ (9 cm). The common name comes from two characteristics. First, the coloration of adults resembles earthworms: pink with prismatic highlights (like oil on water). Second, they are long, thin, and tend to stay buried in the substrate. Males can be rare in some imports. (More than thirty wild-caught animals checked were all female for the pictured stock in one rearing habitat.) Females that have been in captivity for years are able to produce eggs that develop without fertilization (parthenogenesis), but wild and captive-bred stocks are usually male and female. Females produce 1 mm, yellow eggs year-round when fed well, the damp substrate is a few inches deep, and it contains at least 30% crushed wood and leaves. However, females are not terribly productive, and three dozen eggs a year is good. Individual egg capsules are crumbly and difficult to separate from the substrate. The weak capsules are not necessary for successful gestation. Growing specimens are white until they reach an inch (~2.5 cm), then are a pale pink until the last molt, after about four years. Adults do not molt but can live more than six years.

Orthoporus ornatus (Girard, 1853)
North America's largest native milliped (5-6.5″ or 13-16.5 cm) is one of the most colorful spirostreptidans. Geographically isolated color forms may be found in Arizona, New Mexico, western Texas, and northern Mexico. Adults of the 'pumpkin' color form are banded in brown and yellow and are also commonly called pumpkin stripe or tiger millipeds. This banded form is found in Texas and New Mexico. A reddish-brown

Macrolenostreptus brachycerus
Adults
These adults are alive in 2023, three generations removed from Tanzanian imports acquired from International Dragon Breeders Inc. in 2002.

Macrolenostreptus brachycerus
Adults and immatures
Immatures are a pale white-pink until the ultimate molt.

Orthoporus sp.
Texas

Orthoporus from Arizona and chocolate brown form from Texas are commonly available—they may also be *O. ornatus,* but members of this genus have to be dissected and identified by the gonopods. *Orthoporus flavior* is found in Texas and Mexico. There are an additional eighty species named from Mexico down through South America, but the majority may end up junior synonyms.

Orthoporus require high humidity like other millipeds, but at the same time should be provided more ventilation or a tall enclosure without deep substrate. Extreme dampness and condensation should be avoided. Over the long term these conflicting requirements can be difficult to consistently provide in captivity. However, wild adults usually live one to three years in captivity, which is probably the natural longevity. The sticking point in reproduction, as with many spirostreptidans, is convincing the adults to lay eggs in artificial habitats. Internet reports of breeding to this point have all been pest julidans mistaken for baby *Orthoporus.* I am only aware of one person who has bred these successfully under artificial conditions (Barney Tomberlin, Hatari Invertebrates, pers. comm.). There have been few, if any, serious attempts to replicate the seasons so it may relate to light, humidity, temperature, or some combination. Nevertheless, convincing females to lay eggs is difficult by any standard. Tomberlin's success employed vats with deep substrate in an enclosed awning strongly influenced by natural parameters in southern Arizona.

Orthoporus ornatus "Pumpkin"
Texas

Orthoporus texicolens Chamberlin, 1838
This handsome *Orthoporus* is usually dark brown with metallic margins, but there is a lighter, banded form. Specimens hail from the Gulf coastal plain of Texas and northern Mexico. The handsome adults are like other commonly kept *Orthoporus,* except the adult genders show extreme size dimorphism. Observed males of the depicted stock are considerably smaller than females, barely half the mass. Males of other *Orthoporus* species tend to be smaller than corresponding females, but the upper and lower size ranges of the genders overlap. Adults live at least three years in captivity, but the impetus for egg laying is unknown.

African Olive
Plethocrossus sp.
This large African milliped, popular in 1990s imports from Tanzania, is deep olive green with golden legs. Adults are thick-bodied and reach 6″ (15 cm). This species has similar body proportions to *Archispirostreptus,* except the adult males have a visible hood expansion of the front seven segments. Egg laying in captivity has occurred early fall into late winter. Eggs are large, 2 mm diameter, and yellowish. Each egg is laid in a very large, triangular case, which often has fibers and pieces of substrate attached. Exposed eggs are extremely sensitive, which is unusual for spirostreptidans, so it is important not to damage capsules. Adults can be removed from the enclosure once a number of capsules are present, since they can damage the capsule. Development through

Orthoporus ornatus
Arizona

Orthoporus ornatus "Pumpkin"
Texas
Nibbling on finger

Orthoporus texicolens
Mating (the male is the little one)

the early instars is easy with moderate die-off. Early stadia chew through rotten wood and leaf substrates, but feed almost entirely on dog food and fruit as they grow larger.

Congo Black
Spiropoeus fischeri Brandt, 1833
Congo blacks are found throughout central and west Africa, from the Congo, Central African Republic, west through Gabon, Ivory Coast, Benin, and Guinea. Animals were traded under the junior synonym *Mardonius parilis acuticonis* for many years. They were also traded under the '*Archispirostreptus gigas*' label and accounted for a disproportionately high percentage of historical *A. gigas* reproduction reports. They have been relatively rare imports but reproduce more readily. Congo black millipeds have been bred and traded in the United States since 1999 (Harrell 2005) but probably were around decades longer and simply mistaken for AGBs. Development takes up to five years for young to reach adulthood in captivity. Reproduction is common in captivity, but adult females can wait more than three years following the ultimate molt before mating and laying eggs. Despite three generations of data, the adult lifespan is uncertain. They seem to die within six months of reproduction, about three years following the ultimate molt, but can die as little as a year after maturity if conditions lead to earlier reproduction. Ova are laid in a large pile buried at the bottom of the substrate, often against the glass bottom of the enclosure. The 1.5 mm eggs are tiny compared to those of *A. gigas* (and *A. gigas* lay each egg separately). Early instar specimens can be difficult to discern from *A. gigas*, but older animals have a satiny ridge along each segment and the head is smaller in relation to the body. Captive specimens grow better and do not have the minor segment deformities common on cultured AGBs. Captive-bred specimens normally have much darker legs than *Archispirostreptus* reared under the same conditions. No commensal mites have been reported, but most specimens available have been captive reared. Body length for the male is around 5.75″ (14.6 cm) while females are 6.5-7″ (17-18 cm). Specimens are better for handling than AGBs since they almost never exude defensive secretions or feces on the handler, and rarely curl up for long. Subadults are famous for completely cannibalizing exposed tankmates immediately following a surface molt, rarely leaving more than a few legs behind.

Green Pinkleg
Spirostreptus gregorius (Attems, 1914)
This handsome spirostreptidan from Angola and Congo can exceed 5″ (13 cm) in length, though it has a very narrow, wiry body. The trunk is light army-green with dark margins at the back of each segment. The pale-pinkish legs are longer than the diameter for the body, and even longer on the adult male. It seems built to be arboreal, but branches placed in the captive habitat do not get much use. Like a few other species, the male's legs pulsate rhythmically during courtship and mating. Immatures feed on a variety of items and take four to six years to reach maturity. The ultimate molt can take as long as two years. Adults live six years or more and do not molt.

Fire-Head
Spirostreptus serviatus Attems, 1914
The fire-head is a handsome, bi-colored, large species from west and central Africa

Plethocrossus sp.
Tanzania
There is a mating pair, while a second male tries to horn in.

Spirostreptus sp.
Subadults in the double-coil defensive position that is often observed for this species
Stock from Tanzania (circa 1995)

Spirostreptus gregorius
Late instar immatures

that is only available as captive-bred offspring. Wild adults are rarely seen, even from European vendors where it is commonly maintained in breeding. Like many of the larger spirostreptidans, raising small specimens to maturity lacks significant challenge, but getting adults to start laying eggs is difficult. These prefer warmer temperatures from 75-85° F (24-29° C). At below room temperature, specifically 68° F (20° C), they become limp, but have been able to survive a few months this way. A 2″ (5 cm) specimen will take at least three years, if not five, to reach maturity. Even very large immatures stay buried in deep substrate and do eat a lot. They will surface for watermelon and some other fruits. Most mature females are around 8″ (~20 cm) and males are a few cm shorter. When mating is observed, eggs should follow within weeks. If no eggs are seen, set up a new enclosure with ground leaves and well-decayed wood and clear of all pests. Once egg laying has started, each female can lay 500 or more eggs over a six-month period.

Spirostreptus gregorius

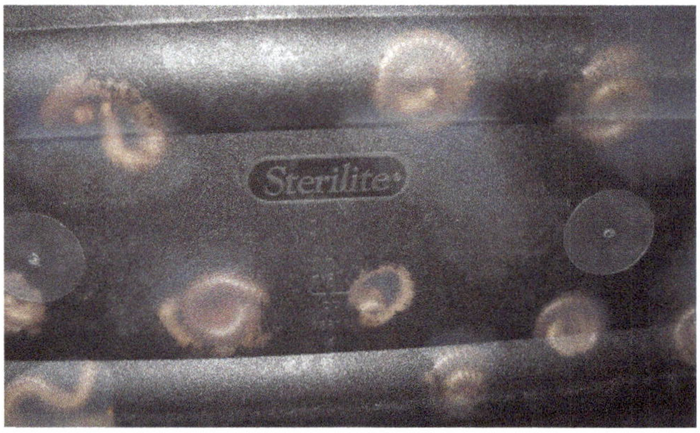

Spirostreptus sp.
Molt sequencing in Feb. 2021

Tanzanian Pumpkin Green
Spirostreptus sp.

This species has been listed as 'globular,' 'Spirostreptus type 1,' and 'Burmese.' The last one is a mystery since it is native to Tanzania (Africa), not Asia. Females grow to 6.3″ (16 cm), while males average fifteen percent shorter. 'Globular' refers to the way the animals form a defensive coil in a three-dimensional glob rather than the usual flat spiral. This is one of the oldest continuously-maintained stocks of captive-bred spirostreptidans. The 2023 specimens pictured on the back cover possibly date back to 1995 stock. One unique feature is that larger immatures can have patchy

Spirostreptus serviatus
Mature adults

Spirostreptus hamatus
Adults
Tanzania

Spirostreptus serviatus
Captive-bred immatures

Spirostreptus sp.
Babies

Telodeinopus aoutii
Late instar

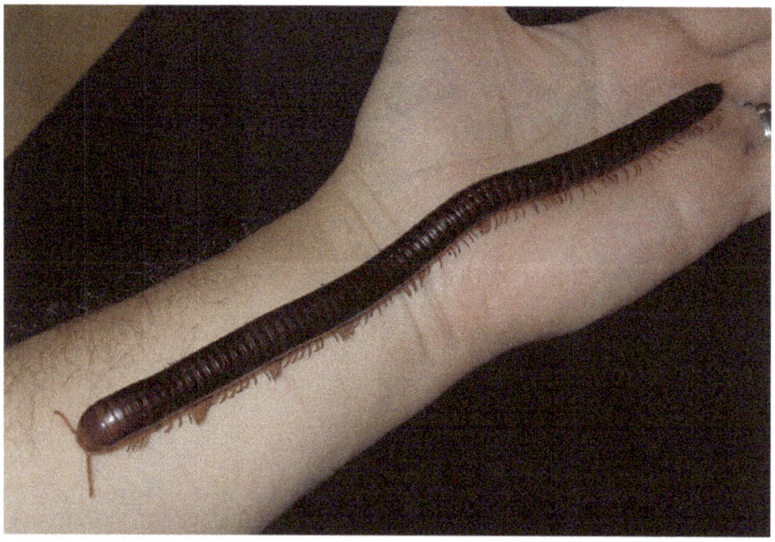

Thyropygus allevatus
Adult male

Thyropygus sp.
Thailand import

bands of light-colored segments. This is not known for any other species so far. The irregular patches darken as specimens mature, and while a faint trace can carry over to adulthood, it is barely noticeable. The survival rate of the immatures is 90% or better unless starved or severely overcrowded. They take time to reach adulthood, three to five years, and an inability to get mature females to lay eggs is often reported. However, if and when a female does finally lay eggs, she produces more than five hundred over a six-month period. Eggs are laid separately, usually in tiny cavities, but not capsules easily separated from the substrate. Heavy nematode infestations can eat eggs before they develop and the empty eggshells are somewhat cryptic. However, lack of egg laying is mostly likely due to the depth and composition of the substrate. Females that oviposit in poor conditions or during the wrong season are not at an evolutionary advantage.

Giant Stripeleg
Telodeinopus aoutii (Demange, 1971)
Named after an Ivory Coast milliped researcher, Akossa Aouti, this large, up to 7″ (18 cm), handsome species boasts striped legs that are more spectacular in real life than in photographs. The giant stripeleg's native range is Ghana and Togo (Africa). It is another captive-bred species that never made its way over among wild imports before the ban, but a few captive-bred specimens trickled over to North American keepers. In one case from early 2000, a dozen captive-hatched 0.5″ (~1 cm) immatures were obtained, but only two reached maturity (both males) after four years. Deaths were spaced out over the years and none seemed to have an obvious cause. They were kept warm, 75-82° F (24-28° C), which may have had a hand in the die-off. Imported adults tend to die within months of laying eggs (P. Grabowitz, pers. comm. 2011), but there do not appear to be any reports for captive-bred longevity. The two males reared from the year 2000 captive-hatched offspring lived two years following the ultimate molt.

Siamese Pointy Tail
Thyropygus allevatus (Karsch, 1881)
Siam is the old name for Thailand and 'pointy' refers to the telson spike. The telson spike is similar looking for many members of the Family Harpagophoridae, a big group of large-bodied, mostly Asian and Indian species. Unfortunately, there has been very little success reported for laboratory reproduction of members of this family. The Siamese pointy tail can be a very heavy feeder that elongates greatly after feeding on pumpkin or watermelon, but also shrinks back after some time without feeding. Recently fed adults are 7-9.4″ (18-24 cm). Healthy specimens usually do not exude defensive fluids during handling, but damaged and dying imports often do. This species is sometimes found with large, brown commensal mites that have elongate front legs like one of the mite species found on *Acladocricus*.

Hot Pinkleg
Tropostreptus hamatus (Demange, 1977)
This moderately large (rarely over 4″ or 10 cm) species was a common import from Tanzania (Africa) from the late 1990s to 2006 in the U.S. and continues to make its way to European specimen collections. The type specimen is from Tanzania, and may be restricted to that country. Adults

are brown with a gold sheen on the segment joints. The coloration would appear relatively bland if not for the hot pink legs. Eggs are laid in individual capsules, but they are crumbly structures that are difficult to differentiate from the surrounding substrate. From hatching, specimens take three years or more to reach maturity under captive conditions. Wild-caught adults rarely live two years. Although easily reproduced, captive-hatched never showed up on vendor lists.

(Top Left) *Thyropygus allevatus*, face
Thailand

(Middle Left) *Thyropygus allevatus*, telson
Wild-caught specimen with a few missing legs and damage marks on the body rings and valves

(Bottom Left) *Thyropygus* sp., telson
Thailand import

(Above) *Thyropygus allevatus*
Thailand
Len Worthington (CC BY-SA 2.0)

Order Spirobolida

Philippine Giant Blue

Acladocricus sp. Ex. Negros Island, Philippines

Negros Island is the fourth largest island in the Philippines, complete with forests, mountains, cities, and croplands. To get an idea of the size of this island, consider its human population is 4.4 million. The species depicted here was collected from Negros Island in 2005. Since only a few percent of the island's original rainforest remains, the original specimens may have been eking out an existence in a small area of regrowth or a tree island between two plots of farmland. Considering the increasing rate of human land use, there is a good chance the habitat and original bloodline of this species have been or will soon be obliterated. The bloodline pictured and discussed here has continued in captivity from 2005 to 2023 in good numbers.

While the genus of this animal is known, the species remains uncertain. There are a number of similar looking species found in the Philippines including *Acladocricus cupulifer*, *A. major*, *A. philippinus*, and *A. porus*, but any specific labeling would just be a guess. There are around twenty species in the genus *Acladocricus*, many on different islands, from Southeast Asia to Australia. The genus comes from the Greek *klados* meaning 'a shoot or young branch,' the prefix *A-* means without, and *krikos* means 'circle or ring,' so the name refers to the telson ring's lack of a spine. This feature is usually standard for the Rhinocricidae (as the family name means 'horned telson').

Other familiar rhinocricids are the small-bodied *Anadenobolus* from the Caribbean and massive *Salpidobolus* from New Guinea. This family is found across both the Neotropics and Malesia (islands between Asia and Australia, including Indonesia, Philippines, and New Guinea).

Captive-reared animals are sexually mature at 4.8-6″ (122-152 mm) but they may grow larger in nature. Adult specimens are approximately 0.5″ (11-12 mm) in diameter. This is the largest spirobolidan with a long-term history of captive breeding.

The giant blue is often a rather spectacular blue color, but only in the adult stage. Even for mature animals, the brightness of the blue is relative to age, condition, and proximity to a molt. Specimens in the early instars are pale and transform to gray overall after three or four molts. Large immatures are brown with gray or green overtones. Subadults are usually army green overall. Only mature specimens are blue, but they can change to greenish or grayish if a molt is imminent.

Acladocricus sp.
Negros Island, Philippines

Acladocricus sp. rarely maintains a defensive coil for more than thirty seconds during handling. They are also unlikely to poop on the handler. Long-term coiling and pooping are signature moves of the AGB (*Archispirostreptus gigas*) which may add to these offenses with a release of defensive chemicals. Sigling (2010) says this *Acladocricus* displays "no release of defensive excretions." Wild-caught adults seldom release defensive fluids and short-term experience would suggest a lack of this ability. However, the adult *Acladocricus* exude copious amounts of defensive fluid when disturbed during or immediately after a molt. Dying specimens can also exude the potent fluids when handled. Immature specimens usually do not release the fluid when picked up but will defend themselves if repeatedly touched (as may be required for definitive gender determination).

Development and Generations: Each ovum is laid individually within an egg capsule. The capsules are unusual because they are extremely difficult to differentiate from frass. Spirobolidan egg capsules and frass usually look different in shape, color, or texture. It may be impossible to identify capsules from *Acladocricus* without opening them. Many eggs seem to be laid bare, though it is possible the capsules are accidentally damaged by the adults. Exposed eggs normally survive early development (unless dried), suggesting adaptation to weak or missing capsules. For most spirobolidans, if the capsule is opened or damaged the exposed eggs' survival rate is greatly reduced or eliminated.

The first captive-reared generation seemed to be very difficult to breed and the adult lifespan maxed out at four years. Males lived half that long. Captive-reared adults produced few, if any eggs, even with a quality, ground-up, rotten wood substrate preferred by this species. Females may only be able to retain sperm from mating for weeks. Recent generations have laid eggs on and off for eighteen to thirty-six months, and three females produced more than 200 larvae. The last male died at about four years of age and production of eggs seems to have stopped.

Immatures are easy to rear as they can survive poor conditions and lack of food better than other captive spirobolidans. Still, good base substrate greatly improves the survival rate. Specimens take about four years to reach maturity. Adults live another two to six years. The first generation of captive-reared adults succumbed to fungal infections within three years of maturity at most. Few, if any, males lived long enough to encounter this problem. Fungal infections were less common in cold 70° F (21° C) and wet conditions. Adults molt twice a year and some would molt out of the fungus, for a while at least. Subsequent generations have so far shown no sign of the 'old-age' fungal infections despite many living longer than the first generation.

Wild adult specimens often host up to a dozen large commensal mites. The female mite living on the species discussed here has extremely long front legs held forward like antennae. A new mite species, *Scissuralaelaps huberi*, was described from Philippine *Acladocricus* sp. (Seeman and Alberti 2015). The adult male *S. huberi* resembles the male mite found on this species, but the female *S. huberi* has short front legs. Unfortunately, in captivity it has not been possible to keep the mites alive for much longer than six months, let alone across generations. Other milliped commensal mites are

(Top to Bottom)
Acladocricus sp.
Negros Island, Philippines
Different life stages and colors

(Top and Bottom)
Acladocricus sp.
Negros Island, Philippines
Different stages and colors

normally lost in culture after a few years or longer—the long development time for immature millipedes means they have no adult host to transfer to.

Enclosures: The short legs and large body suggest these are terrestrial in nature (they should have difficulty scaling trees or rock faces in nature). Still, they can climb the joints of a glass terrarium and should be carefully housed to prevent escape and subsequent rapid desiccation in a home or laboratory. Specimens escape shallow enclosures if the lid is not tight or if there are vent holes as wide as the diameter of immatures. If they find the substrate nutritional, they may not come to the surface for months. Small immatures can stay buried for the first nine months of life if the substrate composition is adequate and is not consumed. They do well on a diet of rotten hardwood and dried hardwood leaves, including oak, beech, ash, and elm. Substrate can be mixed with composted manure, but rotten wood alone is successful if well-decayed. Slices of cantaloupe, watermelon, and pumpkin serve as occasional treats.

I have kept multiple generations successfully in the usual range of 22-24° C (72-75° F). Some have been kept as low as 19° C (67° F) for about three months without noted problems. They grow more quickly toward the high end of the range, but they tend to eat up the substrate quickly and are more likely to encounter damaging molts related to malnutrition. Sigling (2010) states that 26° C (79° F) is optimum; 26° C also happens to be the average daytime temperature on Negros Island. Specimens are among the few tropicals capable of surviving near freezing (not freezing) temperatures for short periods. When I originally acquired a dozen 1″ captive-hatched immatures, it was winter of 2005 and the box came in nearly frozen. All twelve *Trigoniulus macropygus* in the shipment were dead, but all twelve *Acladocricus* survived. The *Acladocricus* did not move until after they warmed up.

With very few available species in North America today, *Acladocricus* is the largest species other than the rarely seen *A. gigas* with any sort of availability. Unlike AGBs, which are occasionally imported through supply companies, the *Acladocricus* seen today in North American captive stocks date back to imports before the enforcement of the importation ban that began in 2006. If this is an endemic Philippine species, it seems likely it will be (possibly already is) wiped out with wholesale loss of habitat. Hopefully dedicated zoologists will keep this species thriving in captive conditions indefinitely. However, indefinite culture is currently unlikely. I am only certain of two successful breeders. This species is among the easiest to culture, but there is time involved and there are very few dedicated, long-term milliped keepers. While it is unfortunate that the unusual commensal mites could not be maintained in captivity long-term, at least the fungal disease affecting mature specimens also died out in captivity more than fifteen years ago.

Puerto Rican White
Anadenobolus arboreus arboreus (DeSaussure, 1859)

This island species is similar in size to *Anadenobolus a. gundlachi* and *A. modestior* from Puerto Rico, but it is the largest variety of *Anadenobolus* kept to date, up to 4.5″ (11 cm). The named subspecies occurs in Puerto Rico, British and U.S. Virgin Islands, and some small islets in between. Mature Puerto Rican specimens are chalky, pale gray overall with black legs and black margins on the pseudosegment. They produce tiny (barely 1 mm) pearly white ova that look similar to *Narceus* eggs; however females do not construct the order's hallmark strong and protective egg capsules. Ova are laid independently of substrate quality or consistency. Of course, in poor substrate the young starve to death soon after attaining first instar. Young can be reared to adulthood on a primarily rotten wood diet, but die-off is very high if the wood is low quality or mostly converted to frass. Rotten wood should be brown to light brown and easily crushed by hand. Decaying branches from black locust trees (*Robinia pseudoacacia*)

Anadenobolus arboreus arboreus
Adult

Anadenobolus arboreus gundlachi
Puerto Rico

Anadenobolus arboreus
St. John, Virgin Islands

and related trees are a valuable food to add to the enclosure. Without adequately decayed substrate, over six to twelve months the die-off of immatures will eventually reach 100%. The immatures are very sensitive to pests. Lastly, temperatures below 72° F (22° C) greatly increase die-off rate. Adults can molt after breeding, but few live long enough.

Virgin Island Black
Anadenobolus arboreus arboreus (DeSaussure, 1859)

Although Virgin Island specimens look quite different from Puerto Rican animals, they are considered the same subspecies. The species was originally described by Linnaeus in 1859 from specimens collected on the island of St. Thomas in the Virgin Islands. They were named *arboreus* (meaning "of or from the tree") because the adults are commonly encountered climbing tree trunks. The local term in the Virgin Islands for millipedes is 'gongolo' and photos are sometimes seen with *gongolos* accidentally used as a species name. The Virgin Islands' form was redescribed as *Rhinocricus gisleni* (Verhoeff, 1941) but synonymized (Hoffman 1999) with *A. a. arboreus*. This form tends to be smaller—adult males barely reach 3″ (~7 cm) and females around 3.5″ (9 cm). The mature animals depicted were collected in Maho Bay, St. John Island, following a rainy night in July. The temperature was 80° F and the relative humidity was 71%; collection time was 12:30-3:00 p.m. Twenty-six individuals were found on branches and under the bark of rotting seagrape trees (*Coccoloba uvifera*) (W. Smith, pers. comm.).

Substrate was compost mixed with expanded oak pellets, overlaid with multiple decaying black locust branches. Fruits and dog food were offered and sometimes nibbled on. Eggs were first observed in October. Surprisingly for this genus, there was evidence of a tiny, weak capsule, without space between the wall and ovum. By January 2023 a hundred offspring had grown to 0.3″ (8 mm). In the fifth stadium they are surprisingly dark like the adults—nearly all other spirobolidans and most diplopods are colorless at this stage. Adults stopped laying eggs in November and began post-ultimate molts in late January.

Puerto Rican Orange-Spotted
Anadenobolus arboreus gundlachi (Karsch, 1881)

This beautiful subspecies adds a dorsal row of orange spots to the color of the Puerto Rican *A. arboreus arboreus*. Females tend to be very productive in captivity, but the die-off during early development can be very high. Adults grow up to 4″ (< 10 cm) and obtain maturity in as little as a year and a half. Eggs are easily collected from wild-caught adults and young grow well in the early stadia. Rearing of immatures can be problematic. Die-off is much higher and sooner without regular replacement of good quality rotten-wood-flake soil. The rotten-wood substrate is consumed quickly and should be replaced before more than 75% has been converted to frass. Although a good percent of young can be reared to large immatures or adults, die-off is usually highest in the late stadia.

Hybrid *A. a. gundlachi* x *A. modestior*

In 2011, I acquired a group of barely-visible, 0.25″ long (0.6 cm), captive-hatched specimens from a myriapod breeder who had kept Puerto Rican *Anadenobolus* since 1994.

Anadenobolus arboreus "Red"
Puerto Rico

This time he kept his reproductive adult *A. a. gundlachi* and *Anadenobolus modestior* in the same breeding cage, assuming they were incompatible breeders since they are different species and both were collected in Puerto Rico. As the immatures grew to 1.5" (3.5 cm), I imagined there must have been an intermediate juvenile coloration I did not recall, but as they approached mature size the truth of hybridization became evident. The animals pictured here reached 3" (7.6 cm) in less than a year and very much appear to be hybrids of the two enclosure inhabitants. Unfortunately, they maintained the slow, late-instar losses common to *A. arboreus* and displayed no hybrid vigor.

Anadenobolus arboreus gundlachi X *A. modestior*
Hybrids at ten months
Varying immature coloration
2-2.5" (5-6 cm)

Anadenobolus arboreus 'Red'
This is another handsome Puerto Rican geographic color variety or subspecies whose offspring showed up on some vendor lists beginning in 2020. Young may be a little heartier than other Puerto Rican species and do not seem to be bothered by a limited number of nematodes in the decayed wood substrate. Slow die-off in late stadia animals is similar to other *A. arboreus*. Time to maturity is eighteen to twenty-four months and they do not lay eggs before reaching approximate full size.

Puerto Rican Giant Bumblebee
Anadenobolus modestior (Silvestri, 1905)
This the sturdiest of the Puerto Rican *Anadenobolus* species and it is only found in Puerto Rico. Only nine months are required for the immatures to reach 2" (5 cm) from hatching. Adults eventually grow to nearly 4" (<10 cm). For best results the food should consist of well-decayed hardwood with regular offerings of fruit such as slices of cantaloupe and watermelon. Growing animals can require a lot of substrate replacements, as they have a hearty appetite for soft, rotted wood. Like other *Anadenobolus*, cool temperatures increase die-off, but so do temperatures much greater than 80° F (27° C). Wild adults, even those that molt in maturity, do not seem to live more than a year. Immatures grow at greatly staggered rates and reach maturity in eighteen to thirty months.

Bumblebee
Anadenobolus monilicornis (Porat, 1876)
This species is found widely across the New World tropics, both as native and introduced by man. According to Ovalle (2002) this species was introduced to Miami, Florida, from the West Indies or South America in potted ornamental plants. The article goes on to explain that capuchin and owl monkeys at the local zoo bite the millipeds and rub them on their fur to ward off insects and catch a buzz from the defensive secretion. As this species has become very

Anadenobolus arboreus "Red"
Puerto Rico

Anadenobolus modestior
In this species immatures approximate adult coloration after just a few molts.

Anadenobolus monilicornis
Adults and immatures
Miami-Dade Co., Florida (2004)

common in Florida, collectors send them to vendors, so the bumblebee milliped has become a regular at shows and some larger pet shops. The specimens depicted here are from stock originally collected in Miami, Florida, in 2004. As of 2022 they have a strong presence in Orlando urban sprawl and disturbed areas, mostly in mulched flower beds.

The common name comes from body segments alternating in black and yellow like a bumblebee—there is also a cute little 'stinger' at the end. The telson spine is harmless and lacks an actual stinger. Segments pale towards the underside by the salmon-colored legs. Adults reach only 2.25" (5.7 cm), but are thick bodied and are many times the mass of a julidan milliped the same length. Females can deposit <1 mm eggs year-round, an inch or two deep in the substrate. Substrate of coconut fiber mixed with leaves and rotten wood works well for the fragile baby millipeds. Good quality, rotten-wood-flake soil works even better. Barely nutritious substrates and worm infestations can rapidly kill large numbers of growing specimens, adults a little more slowly. Melon rinds are a beloved treat but are not enough alone to keep *A. monilicornis* healthy without a good decomposed wood-based substrate. The young can reach sexual maturity and start laying eggs at 1" (2.5 cm) after just twelve months, but it takes three to four years for females to grow to full size. Adults can live seven or more years and molt every year or two. Die-off can be extremely high if they are kept below 72° F (22° C) or whenever substrate has been mostly converted to frass.

Anadenobolus monilicornis
Adults mating
Orange Co., Florida (2022)

Anadenobolus ocraceus
Canal Zone, Panama

Panamanian Banded
Anadenobolus ocraceus (Brolemann, 1900)
This small Panamanian *Anadenobolus*, endemic to the Canal Zone, has become somewhat commonly cultured in the last few years. It is the only *Anadenobolus* listed here that is not especially sensitive to food quality. This may be because their appetite is strangely subdued. Even when there are larger numbers, the substrate takes a year, instead of weeks or months, to be converted to frass. Specimens barely eat anything and remain buried in edible substrate nearly all the time. Immatures are white for at least

Anadenobolus ocraceus
Canal Zone, Panama

the first six molts, while adults are a beautiful, white-and-brown banded creature. Suggested substrate is finely-ground, decayed hardwood. Fruit and dog food are usually ignored, except when there are hundreds, and the substrate is mostly depleted. Like other *Anadenobolus*, reproduction is year-round and large numbers of young are produced at various stages of growth. They mature at a small size (around 1″ or 2.5 cm) at twelve to eighteen months like *A. moniliicornis,* and continue to molt and grow larger. Specimens take another year to reach about 1.6″ (4 cm).

Thai Rainbow
Apeuthes sp.
This appealing animal has been exclusively available as captive-bred specimens since the late 1990s in North America. The stock depicted in the accompanying photographs was acquired as large immature (five-centimeter-long) captive-hatched specimens in 1997 from a milliped breeder in New York. The parents of these were acquired by the New York breeder from central Europe as small offspring a few years prior. The original adult specimens were collected from an unspecified location in Thailand, an indeterminate number of years prior.

In the U.S. during the 1960s and 1970s, imports of large, colorful, tropical species were a common but sporadic sight. Most imports were the AGB. An increasing variety of regularly-available, imported tropical species (mostly from Africa and Southeast Asia) became available to enthusiasts and zoologists in the United States through the mid-1980s to the early 2000s. In the past, wild-caught specimens of the Thai rainbow were not among the popular species on vendor lists, nor was this species seen at pet stores. Importation of exotic diplopod species for the U.S. pet trade was discontinued in 2006. Because available specimens of this *Apeuthes* were never wild-caught animals, moderate availability of captive-produced specimens has remained through to the present day.

Apeuthes sp. is more colorful than the average spirobolidan, but the common name originated with a different spirobolidan that had been very common at pet stores and was thought to be a close relative in the 1990s. The other rainbow is the recently reassigned 'Vietnamese rainbow,' *Atopochetus (Tonkinbolus) dollfusii* (Pimvichai et al. 2018). The adult *A. dollfusii* does not have a full range of colors like a meteorological rainbow, but there is a red stripe running down the center of the dorsal body surface that fades towards the edges. The red stripe is solid in the middle and where it meets the light gray coloration on each side, the edges resemble the prismatic fading of a rainbow.

The Thai rainbow (*Apeuthes* sp.) has long been mislabeled by hobbyists in Europe (Sigling 2010) as *Tonkinbolus caudulanus*, but specimens were identified in 2003 by a milliped taxonomist, the late Dr. Richard Hoffman, who diagnosed the specimens to the genus *Apeuthes*, not *Tonkinbolus*. Hoffman had been working with the genus *Tonkinbolus* since at least 1962, was intimately familiar with related species, and is co-author of the genus *Apeuthes* (Hoffman and Keeton 1960). Without a more specific collection location, Hoffman said a species identification was not possible. Despite widespread knowledge of Hoffman's direct identification, some hobbyists prefer the feeling of accuracy offered by a complete binomial without concern for the greater inaccuracy created by using the wrong species and genus (and wrong family). To add to the confusion of the labeling, the 'Vietnamese rainbow' *A. dollfusii* also inhabits Thailand (Enghoff 2005) and is concurrently traded under the junior synonyms *Aulacobolus rubropunctatus* or *Tonkinbolus dollfusi*.

The word *Apeuthes* in Latin is a nickname for the Roman goddess of fortune and as a verb means 'to seek to understand.' The genus *Apeuthes* currently contains three species, all described from type specimens collected in Vietnam. Since Cambodia and Laos lie directly between Thailand and southern Vietnam, *Apeuthes* sp. and *Atopochetus dollfusii* are likely found across all four countries, even if they do not range more widely across southeast Asia.

Apeuthes displays emblematic features of the Family Trigoniulidae from egg to adult. The early reproductive stages of the Vietnamese rainbow and Thai rainbow are significantly different in size even though the egg capsules are similar in measurement (McMonigle 2012). The large egg and protonymph size of *Atopochetus* correspond with those of other pachybolids (i.e. *Epibolus* and *Pelmatojulus*) while the proportionately small egg size of *Apeuthes* belies its trigoniulid origin. Mature male *Apeuthes* have similar leg structure to the female, whereas *Atopochetus* males have enlarged legs with ventral pads of setae (common for Pachybolidae). *Apeuthes* have a thinner body relative to the legs that narrows from segments six through ten, which is also common for trigoniulids. The male *Apeuthes* drums its long legs across the female's body during courtship. This is a common trigoniulid behavior, but it can be found in species from both families, and at least one other order (Spirostreptida).

Other commonly kept trigoniulids are *Trigoniulus macropygus* from the Philippines and *T. corallinus* from Thailand and Myanmar. The second is small and relatively plain. It has gained some popularity in captivity in North America because it is an overly-common, easily-collected Florida adventive introduced with landscape plants.

The adult male is the more brightly colored of the sexes, though the overall pattern is similar. The body segments are bluish gray with red margins. The ozopores are marked in black. The legs and antennae are red. In adult female *Apeuthes* sp., the red-margined, steely-blue segments have a blonde band at the front. This band is only visible where the segments bend, when the animal has eaten well. The telson ends in a pronounced, rounded spike that is slightly indented towards the base on the adult male. For immatures and adult females, the spike narrows evenly from front to back, so the telson appears triangular from above.

Specimens can produce strong defensive fluids (released from the ozopores) if sufficiently stressed. If they are handled regularly, they are less likely to release defensive secretions during handling. The fluid contains quinones that stain skin dark purple, then brown for a few days. It causes no sensation on unbroken skin, but stings if it enters cuts, eyes, or mucus membranes.

Though easy to keep and breed, the availability of *Apeuthes* specimens is kept in check by a long life cycle, and to some extent by limited demand. It could be produced in quantity if there was a large market, but it takes some time. Immatures are seldom longer than an inch after a year and many molts. Approximately three years are required from egg laying to maturity. Even after they are sexually mature, it can take an additional two or three years to get good egg production out of a mature female. In one group (9 male, 11 female) one young adult female made hundreds of empty egg capsules eighteen months after reaching maturity, while another laid a few dozen viable eggs. The remaining adult females mated but did not produce eggs for the first twenty-four months (McMonigle 2007). Eggs are usually laid over a period of thirty to sixty days, once a year, and one female can lay hundreds of eggs each season. In captivity the season for egg laying does not correspond to specific months. Eggs may be laid year-round under good conditions, in a large terrarium, containing multiple females.

Adults are normally 3.5-4″ (8.3-11 cm) in body length, with males on the low end. They continue to molt after maturity and a large, older female can grow to 5″ (12.7 cm). A female is unlikely to grow to maximum size in less than five years from hatching with the described food and temperature at around 74° F (23° C). Adults rarely live less than five years. The oldest documented specimen so far is a female that lived 14 years (McMonigle 2012). She was acquired as part of group of ten large immatures (approx. 5 cm in length) in 1997. Six of the ten were still alive in 2009. Two survived through mid-2011 and the last lived into the early months of 2013. The longest-lived female (and her siblings) may have been two years old on acquisition, so the total lifespan of the longest-lived female was, at most, 17 years from hatching to death.

Apeuthes are arboreal millipedes as evidenced by the relatively long legs. Branches can be added to the enclosure for climbing observation, but the animals do not need to climb and do not require tall enclosures. This is one of the few milliped species where you can turn your hand upside-down and shake it and the animal stays put. (I have tested this with adult males but made no attempt to dislodge them with effort.) The adults especially are excellent climbers, so it is very important the lid of the enclosure fits tightly. They cannot climb smooth, vertical surfaces, but a thin film of dirt or sharp corner allows for climbing. Removal of the bead of silicone from the corners of glass aquariums reduces observed climbing but has little effect on keeping specimens contained. This species is certainly found high on trees and rock faces in nature at times, and escapees may be found hanging onto the wall or the uprights of a terrarium stand. All millipeds, but especially climbers, should be carefully housed to prevent escape, because they rapidly desiccate in the home or laboratory. They require damp substrate and very little ventilation. They escape shallow enclosures if the lid is loose.

They can also escape through vent holes if the holes are the same diameter or larger than the body rings; immatures are often lost this way and dry up much more quickly.

Ground-up rotten hardwood and dried leaves, mixed with composted manure (available from most garden shops) is a successful substrate. However, other composts or decomposing hardwood flakes can be used as singular media if other items are unavailable. Though not necessary, substrate should include decomposing leaves, rotten wood, or both. Hardwood pellets for wood-burning grills can be used in place of rotted wood when mixed with damp substrate. If the substrate is adequately nutritious, they rarely come to the surface. Young specimens can stay buried for months. Dried dog food kibbles, powdered milk, or slices of cantaloupe, watermelon, and pumpkin serve as occasional treats.

Data for breeding and rearing this species has been acquired almost entirely in the lower range of 'room temperature,' from 67-74° F (19-23° C) year-round, with the cooler temperatures in winter. Thailand is a hot country, and the lowest average monthly temperature is a balmy 79° F (26° C). Nevertheless, the temperature of the microhabitat for this species is uncertain. When kept at 75° F (24° C) or warmer, the immatures grow faster, but there is limited data on long-term effects of warm temperatures and long-term results in captivity. Faster does not always equal better.

1st instar *Atopochetus dollfusii* (left)
1st instar *Apeuthes* sp. (right)

The Thai rainbow continues to be bred in captivity and maintains limited regular availability through specialty vendors. It is one of the hardiest species and the documented longevity is unparalleled. It is longer-lived than other commonly maintained species. A number of large spirobolidans and spirostreptidans are known to live three to six years after reaching maturity; nevertheless first-hand longevity documentation is scarce. *Apeuthes* sp. is a fantastic arboreal captive and while not prismatic, it is still a pretty creature.

Madagascan Fire
Aphistogoniulus sp.
Though it was a regular among arthropod displays, availability from Madagascar had been seasonal. The average milliped enthusiast was familiar with this species even if they had never seen one. Much of its fame resulted from it being among the more established imports between 1980 and 2006. Numerous pictures can be found in books and online. Nevertheless, its true fame is due to the adult's fantastic coloration. The vivid red-and-black striped body and yellow legs result in a milliped with unmatched intensity. Size is also a factor since it would not have made a good display if it were tiny. It is a thick-bodied species and a mature female can grow to half a foot in length. Despite prevalence in collections and zoos, and numerous breeding attempts, nobody bred this species the first few decades it was kept.

Apeuthes sp.
Thailand

Aphistogoniulus sp.
Captive-reared specimens mating (2009)

Atopochetus dollfusii
Adults and young
Note the large relative size of the first instar larvae.

The yellow stain and pungent quinone smell is the major component of fire milliped exudate. However, *Aphistogoniulus* secretion must include an additional chemical different from other spirobolidans because the secretion includes a notable and unique fruity smell.

I first kept this species in 1994. I had heard about it for a long time and wanted to try keeping this beautiful animal. Of course I hoped it was possible to reproduce them. Ten specimens were acquired for a small fortune and they lived for a few years, but there were no signs of reproduction. They were kept on quality, aged compost manure and fed mostly leaves, dog food, and a little rotten wood.

A decade later, 2004, another attempt was made. Eight arrived unsexed but (luckily) were discovered to be four males and four females. Adult males are 25-30% smaller than females, so gender determination is straightforward at a glance. Size cannot be used to sex immatures, but it is consistently accurate for adults.

On arrival they were set up in a 1.5-gallon (5.7-liter) plastic shoebox with fourteen 0.0625" holes drilled in the lid for ventilation. The shoebox was filled within a few inches of the top with 2/3 hand-crushed, soft brown rotten wood and 1/3 coconut fiber. The moisture level was maintained so the uppermost layer of substrate never dried. Additional food was offered every week or two: dog food, watermelon, or brown sugar water. Leftover fruit or dog food was removed by the second day. All of them molted without an appreciable size increase within the next six months, but I was not surprised there were no sign of egg production despite the occasional observance of mating pairs.

After a year the substrate had been converted almost entirely to frass and one day a conspicuous, large, oval object appeared on the surface. I have seen many egg capsules from various millipeds and most can be tough to differentiate from the frass even though the surface texture and shape is different. However, this one was obvious right away because the shape, texture, and especially large size were very distinct.

Forty-six ova were laid in four separate batches, but there was no way to know if each batch was laid by a different female, or all by one. Only twenty-five of forty-six eggs hatched. Hatch time was approximately forty days. Despite a high 46% ova die-off, the immature millipeds were very hardy and experienced 0% die-off over several molts.

Batch	1	2	3	4
Date	9/19/05	10/25/05	11/13/05	12/1/05
Eggs	15	10	15	6
Hatch	5	9	8	3

The young millipeds were kept in a damp mix of frass and rotten wood, about 50/50. Each batch was placed in a 16 oz deli cup with a dozen pinholes in the lid and substrate three-quarters to the top. Additional feeding consisted of 1-2 ml of brown sugar water (one part sugar to three parts water) or a tiny pinch of protein powder every few weeks. At seven months the oldest group of five were approximately 2" long (5 cm) and readying for the next molt. The lack of die-off was impressive.

Ten specimens were kept, grew to adulthood, and were first observed mating in late 2008. However, no eggs were laid by 2011 and three adults had died, so they were traded off to be used in a display. Four egg capsules were observed in 2011 when the adults were traded off. All four hatched and grew to 2″ (5.1 cm) by 2012, and were also traded off for a display because they were no males in the group. As with the previous generation there was no immature die-off (of four).

With a correctly vented cage, consistent moisture, and a rotten wood diet (supplemented with the occasional piece of dog food) these hardy millipeds live many years. Overcrowding should be avoided because adult *Aphistogoniulus* molt regularly, about every six months, and need adequate space and substrate for individual molting cells.

Vietnamese Rainbow
Atopochetus (Tonkinbolus) dollfusii (Pocock, 1893)
In the 1980s and 1990s this was among the most common and least expensive species found in North American pet shops. It hails from Vietnam and Thailand. The mature specimen is a handsome diplopod, but it does look nicer in photographs than in person. It is easily bred, has huge eggs, and young grow readily to reach adulthood in thirty months, but it died out of North American collections within five years of the import ban. The primary diet of rotten wood and leaves must be replaced regularly for good development. The problem is they not only require consistent, high-quality food, but also die if they are allowed to go outside a narrow temperature range. I had a large colony that took years to develop and included hundreds of immatures and subadults. One summer the temperature went a hair over 82° F (28° C) for a few days and every last one died. They also die (not as rapidly) below 72° F (22° C). Adults live for a few years, possibly longer if conditions can be kept within a narrow range.

Benoitolus sp. "Khao Sok"
This is a handsome creature with greenish-gray segments, cream bands, and light orange legs. Specimens of all ages glow brightly under ultraviolet light. It may not be coincidental this is a pseudospirobolellid and the other highly fluorescent spirobolidan

Benoitolus sp.
Khao Sok, Thailand.
Coloration under daylight versus ultraviolet light

maintained in captivity is a spirobolellid. Adults can exceed 2″ (>5 cm), but they are thin-bodied and lack appreciative mass. There are two species described from Thailand (*B. birgitae* and *B. siamensis*), but the specimens depicted have not been identified and are labeled Khao Sok. The original adults were collected in or near the Thailand national park of the same name. This is a highly reproductive and easy species, so even in countries where imports are still allowed, specimens on vendor lists are from breeders.

Mozambique Crimson Blackleg
Centrobolus anulatus (Attems, 1934)
Adults are magnificent, shiny red animals with black ozopores, legs, antennae, and a black head. Mature specimens barely reach 2.5″ (6.4 cm) but they are as thick-bodied as a five-inch *Orthoporus*. Immatures start out pale with conspicuous, dark pigmentation around the ozopores. They begin to approximate adult coloration when they reach an inch in length.

This is one of a select number of tropical species not imported from its country of origin into North America, but made its way from European breeders. Specimens originated in South Africa, where they range from northeastern Kwa Zulu-Natal to southern Mozambique. It is rarely seen from U.S. vendors but can be reared with some consistency.

Millipeds with short developmental periods often have a short adult lifespan. The crimson blackleg grows rather quickly and reaches adulthood in around eighteen months. Unfortunately, adults only live about that much longer so the majority of specimens are dead within three years of the eggs being laid. This tropical species should be kept at 75-80° F (23-27° C), but temporary deviations outside this range are not harmful.

Ova are laid in individual egg capsules. The capsules are more difficult to differentiate from frass than those of most species, partly because of the small number created, but also because frass and capsules look similar. The number of eggs laid is in the low dozens (McMonigle 2009) though hundreds ought to be possible. The immature survival rate can be very low in the early instars. Once they have molted a few times, few deaths occur.

Rotten wood is the primary diet although shredded, decayed leaves are almost as well accepted. Rotten wood must be well-rotted and soft in order to be a useful food. Fish food pellets are a welcome treat, but fruits are not eaten with zeal.

Centrobolus splendidus (Attems, 1928)
Adult females of this South African species rarely exceed 3″ (7.6 cm), and males are a little shorter. This is a relatively easy captive species with a preference for pest-free, finely-ground, decayed wood. Supplements including dog food and melon are eaten sparingly. Although there is photographic documentation that *C. splendidus* immatures were hatched and maintained in captivity at least as far back as the late 1990s, captive-bred specimens did not appear on vendor lists prior to 2015. Eggs can be laid any time of year, but usually there are only a few dozen over about a month's time. About six months later the adults molt and some months later may mate and lay again. The first few stadia are extremely sensitive to poor substrate and nematode infestation, but once immature specimens have molted four times they rarely die. Adulthood is

Centrobolus sp.
Mozambique

Chicobolus spinigerus
Adult

Chicobolus spinigerus
Miami

commonly reached in eighteen months and the adult stage is about the same length. Some adults succumb to old age in less than a year, while others last almost two.

Ivory

Chicobolus spinigerus (Wood, 1864)

This is the only described species in the genus. Common names have included 'black & white,' 'Florida ivory,' or just 'ivory.' The name 'black & white' is certainly a more descriptive name, but it lacks the flair and familiarity of 'ivory.' The earliest reference to a common name I have encountered for this species is 'Florida Ivory' in an article by Walls in the first issue of *Reptile Hobbyist,* 1995. The problem with including Florida in the name is that while there are some large spirobolidans restricted to Florida, this species is not one of them. It occurs over a large area of the southeast. It is documented from Alabama, Florida, Georgia, and South Carolina, but I found specimens under the bark of a huge rotten log at Jellystone Campground in Barren County, Kentucky, in 2009, suggesting the natural range is more extensive than commonly reported.

Centrobolus splendidus

Adults and immatures look very different. The first few instars are translucent white. The black dorsal coloration shows up as gray to pink after a number of molts and finally darkens to a near black when the animal is around two inches. Larger, older adults tend to have the greatest contrast between the dark top and light 'belly.' The standard mature coloration is unique and cannot be confused with any other spirobolidan known at this time. Dark forms usually look banded in the middle stadia.

The ivory begins life as a tiny (1 mm), round, yellowish egg. The egg is surrounded by a hollow, thin-walled capsule that is about 3 mm across and an irregular oval shape. The egg is nearly always accompanied by 300-500 siblings, which are laid between April and June. An uncommon aspect of *Chicobolus* capsules is they are created and deposited in a single pile rather than randomly around the enclosure. The female chooses a spot and digs down 2-3" (5-7.6 cm) and sets about creating capsule after capsule. She forms approximately twenty capsules each day and does not move from her chosen location until finished. She will form the mass of capsules much closer to the surface if the substrate is shallow and may continue in another area of the enclosure if disturbed.

Molts slow down with age but sexual maturity is reached in about eighteen months. The majority of spirobolidan species are not sexually mature until they achieve nominal adult size; sexually mature specimens continue to molt but do not grow larger. Ivory millipeds are one of the odd species (like *Anadenobolus monilicornis*) that continues to grow significantly post maturity. Males of the same approximate age are normally about 2/3 the mass of the female but an old male can be a lot bigger than a young mature female. Maximum size is obtained three to four years after adulthood, but some die without post-ultimate molts. The total lifespan can range from three years to a decade.

With decent care, breeding is extremely easy, and this is the best large spirobolidan for a beginning zoologist. *Chicobolus spinigerus* was the first milliped I successfully reproduced, and the conditions were substandard for breeding other millipeds. Egg production is always in the hundreds, even for small, newly-mature females. If fed poorly during the first few instars, the immatures can die in huge numbers. With attention paid to food quality, their survival can exceed 90%.

Ebony
Chicobolus spinigerus [color variation]
This form is the same as the normal phase in every way except color, though populations are geographically isolated and the difference in coloration has been consistent for offspring. The original description was *Chicobolus pilsbryi* from Boca Chica Key (Monroe County, Florida) in 1947, but the designation was synonymized with *C. spinigerus* by Keeton 1960. Specimens from Miami (Miami-Dade County, Florida) can also be dark overall.

Captive reproduction and growth appear to be identical in methods, survival, production, and time required to the common color form. The immature specimens have white underneath and appear darker at the same size, but they cannot be discerned from the ivory form until they reach a few inches in length.

Slate
Comanchelus chihuanus (Chamberlin 1947)
This is an uncommon southwestern genus associated with limestone and algae. *Comanchelus* species range through Arizona, New Mexico, nearby Mexico, and western Texas. There are three nearly identical species across Texas, but they all seem to get labeled with this name. The only thing interesting about these animals is that the second segment is reduced. Wild specimens produce whitish green frass, but will feed on rotten leaves and produce brown frass after a week or two. The few specimens that have made it to captivity have lived just a few weeks (Walls, unknown sp.) to four months (pers. observ., *C. lobatus*). Since they fare poorly and look superficially like a slightly smaller and less impressive *C. spinigerus* 'Ebony,' there is no reason to try to keep members of this genus in captivity.

Giant Redleg
Epibolus pulchripes (Gerstaecker, 1873)
Imported specimens from Kenya and Tanzania were as thick as a full-size *Archispirostreptus,* though the animals were a few

Chicobolus spinigerus "Ebony"
Immatures

Chicobolus spinigerus "Ebony"
Boca Chica Key, Florida

Comanchelus lobatus
Uvalde Co., Texas

Epibolus pulchripes
Male and female adults
Second generation

inches shorter and exceedingly more colorful. This magnificent creature is among the largest spirobolidans, but the captive stock depicted is much smaller than wild-caught specimens from the past. It is uncertain if the size discrepancy is related to age (as it may grow larger in post-ultimate molts), genetics (if the captive stock is a smaller geographic variety), or rearing conditions. Males are glossy black with fire engine-red faces and legs. Upon the molt to maturity the males' legs become huge and are outfitted with thick pads used to hold onto the chocolate-brown, satiny female. The female's dusky pink legs are smaller, normally shaped, and lack adhesive pads.

Females create egg capsules that are a little larger than most other species, but the eggs inside are comparatively monstrous—about four times the mass of *Archispirostreptus gigas* eggs! Hatchlings are very sensitive to low-quality substrates and should be provided a primary diet of well-decayed hardwood. At times they will eat large pieces of watermelon and shredded leaves, but if those are not eaten within a day the food should be removed, replaced with

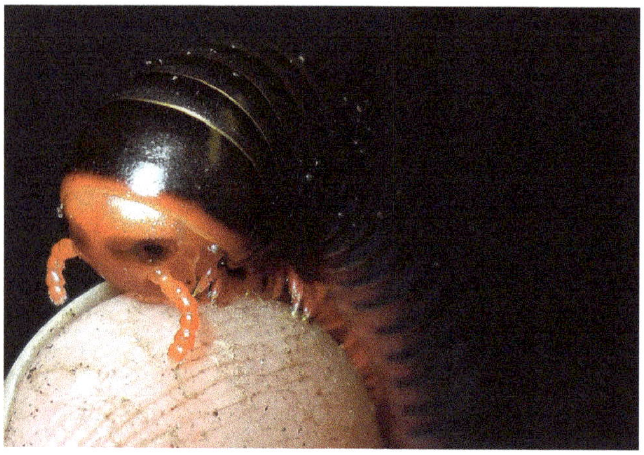

Epibolus pulchripes
Late stage immature male
Adhesive tarsal pads not yet developed

well-decayed substrate, and tried again in a few months. Immature males and females are easy to tell apart at 2″ (5.1 cm) as the surface texture and color begins to differentiate. Despite its huge size, *E. pulchripes* reaches adulthood faster than many smaller giant millipeds, in just two years. Two to three years after reaching adulthood the adults can become prone to fungal infections which invariably lead to death.

Florida Scrub
Floridobolus penneri Causey, 1957
As the name suggests, this species is endemic to Florida and is found in dry, scrub habitat. The shape of the body is unlike other spirobolidans found in North America. It is among our fattest species, half an inch wide, but it is also only a third inch tall, giving it an oval, rather than circular, cross-section. The body shape reminds me of the cloud gate in Chicago. Despite great width, it only grows to about 3.5″ (8 cm) in length. This had been very difficult to acquire, as specimens I obtained in the past were just misidentified *Chicobolus* or *Narceus*. It is only reported from the southern Lake Wales Ridge region in Highland and Polk counties, however the male *Floridobolus* pictured in *Millipeds in Captivity* (2012) was collected among rosemary scrub (the right habitat) in Marion County (the wrong county). I sent the specimen to Dr. Shelley to confirm my identification and it turns out I was wrong. It was not *F. penneri*, but a new species—now *Floridobolus orini* Shelley, 2014. This shortly led to the description of another species, *Floridobolus floydi* Shelley, Phillips, & Bernard, 2014, named from Citrus and Herando Counties. The first two are similar in length, around 3″ (7-8 cm), while *F. floydi* only grows to about 2.4″

Floridobolus floydi
Secretion visible

Floridobolus floydi

Floridobolus orini
Note the width of the body is much greater than the height, resulting in an oval shape to the body segments 11-45.

Floridobolus penneri
Lake Placid, Florida
Immatures

Floridobolus penneri
Molt

Hiltonius hebes
San Diego Co., California
Two females and a male

(6 cm). *Floridobolus* seem to be sturdy creatures. Immatures molt successfully in captivity, living a year or two under general spirobolidan conditions, but the best food and breeding impetus are not yet known.

California Blunt
Hiltonius hebes (Bollman, 1887)
These plump creatures are found from southern California in the U.S., south through Baja California in Mexico. Specimens from the southern portion of the range can have beautiful red bands, but the specimens from San Diego County, CA, are just varying shades of brown. Females can be as wide-bodied as *Narceus gordanus,* but rarely exceed 3.5″ (8.9 cm). The blunt cigar shape allows females of this genus to be easily discerned from *Tylobolus* at a glance. Males are around 3″ (~7.6 cm) with body proportions similar to other North American spirobolidans. Specimens are hardy—wild adults have lived five years in captivity, but have not reproduced. (Keep in mind this longevity and reproduction statement is based on four specimens.) Wild-caught adults molt once a year. These unique creatures were acquired from Hatari invertebrates in 2006, but I have never seen them on a vendor list (only misidentified *Tylobolus).*

Hothouse Whiteleg
Leptogoniulus sorornus (Butler,1876)
This small (~1.5″ or 3.5-4 cm) creature is a pantropical pest and can be encountered in greenhouses anywhere in the world. Easy availability has led to it being maintained in captivity in recent years. There is only one species in the genus. It is related to the commonly kept *Trigoniulus* species, but the survival rate of immatures is much better under poor conditions. Like *Trigoniulus,* they have a relatively fixed mature adult size and do not appear to molt after maturity. Females lay individual eggs over six months to a year and expire without molting. Immatures develop well on crushed, rotten wood or damp, expanded hardwood pellets even in very early stages of decay, and probably any compost or potting soil with decaying cellulose. Large cultures consume melon slices quickly, but if there is a lot of uneaten substrate the fruit is usually ignored.

Eastern Giant Pink
Narceus americanus (Beauvois, 1817)
This is the largest invertebrate in many areas of its range from southern Ohio and Pennsylvania, west to Texas and south into Florida. The body is as thick as a pen and the larger females often grow to 4.5″ (11.4 cm). Males almost never exceed 4″ (10 cm). *Narceus americanus* breed readily once a year, following the winter molt. Females prefer to build capsules out of soft, brown rotten wood. Adults continue to molt post-maturity, usually in the spring prior to egg laying, but adults, males especially, sometimes skip a molt and wait until the following year. The lifespan is short when compared to *N. annularis* and *N. gordanus.* Maturity can be reached in just two years and adults live three to four years after reaching maturity in captivity. Wild-caught adults can die a few months after laying eggs.

Eastern giant pink millipeds are generally absent from urban sprawl and agricultural areas. Even in older forests it can be difficult to find one or two specimens during hours of collecting. Rotten logs or trees pocketed with cracks full of rotting leaves are prime habitat. The size and age of partly rotted trees and logs is more important

Leptogoniulus sorornus

Narceus americanus
Georgia stock
Immatures at eleven months

Narceus americanus
Albino cultivar

than the size of the wooded area. Bear in mind, feeding areas are seldom on the forest floor, but rather are usually one to ten feet (0.3 to 3 meters) above ground level. *Narceus americanus* has recolonized older new growth forests, but occurs in numbers in old growth (often found near or on dead trees three or more feet in diameter). Even small forested areas more than fifty years old are few and far between in the east, let alone old growth. I have observed this species in a number of regions and states, but some collectors never see one because they do not find an old growth area.

Narceus americanus, Albino

Albinism is very rare in invertebrates and many long-time keepers in the last century wondered if it was even possible. There are some white-colored invertebrates including the Puerto Rican white millipeds, small isopods, and spiny flower mantids, but these are not albinos. Cave crayfish lack dark pigments and are not albino either. Nearly all millipeds start out white and many stay that way through numerous instars. Many of the flat millipeds do not develop color until the subadult or adult molt. There are 'blond' forms of species like *Narceus gordanus,* but these still have black legs, eyes, or other areas with dark pigmentation. The *N. americanus* stock pictured here are a true albino variant of a normally colored species. An adult male specimen was sent to a taxonomist, Dr. Richard Hoffman, who confirmed the *N. americanus* species identification.

Narceus americanus usually have reddish margins but in some areas the margins are brownish (Alabama).

The pink bands range from hot pink to barely noticeable, but this coloration seems to be based almost entirely on small differences in food or environment, since related animals in different cages display similar colors (or lack thereof) as a group over time. Usually, the pink is only lost to very poor feeding over a long period. Mature females commonly reach 4.5" (11.4 cm).

Albinism is generally considered a lack of ability to produce the primary dark pigment (melanin in most animals), not an absolute lack of pigment. Like vertebrate albinos, albino *Narceus* are not completely devoid of yellow or red pigments, just dark pigment. Yellow and light red are maintained, but are only notable in older immatures and adults (though the ozopores of small immatures appear red). While some animals can produce a pied or gradient albino, the albinism in *Narceus americanus* for approximately 1,200 individuals produced has been 100%.

The original pair were captive-produced from Louisiana stock and their parents were of normal coloration. It was the only pair produced out of hundreds of young. I received the mature adults in 2004. I had previously reared thousands of *Narceus* from locally-collected specimens and no albino or notable color variation was ever produced. Perry Adkison reported capture of a single wild specimen in Pennsylvania in 2006, but the specimen was lost. Captive reproduction (especially with only one pair) is never a certainty and there was a good chance the adult pair would be the end of their line. Egg capsules were first observed on June 15, 2005. This was a relatively small batch for *Narceus,* as only forty-eight immatures resulted, but it was still a thrilling achievement. Of those, twelve turned out to be normal coloration and thirty-six albinos. The second batch of egg capsules were formed in late June 2006. Approximately 450 immatures resulted from this batch. Only two were normal coloration, and 448 were albinos.

After the very first batch resulted in 25% normal, 75% albino, it seemed the gene for albino was somehow dominant, which did not make sense. However, the second group resulted in just 0.4% normal. The female was housed with normal males prior to acquisition, but had molted since, and it had been more than a year. Every subsequent generation through summer 2022 produced 100% albino young. The most likely conclusion from these numbers is sperm retention in the female can last through molts, and the albino trait is in fact recessive (milliped sperm being last in, first out). 100% of the subsequent generations resulted in albino because females were never exposed to normal males. No subsequent broods of any offspring ever produced a normal color specimen, so the 2005 and 2006 oddities were a result of sperm storage as the female was exposed to normal males prior to my receipt of her.

The original female moted once each year around April, while the male did not undergo any post-ultimate molts in my care. (*Narceus* males commonly go through post-maturity molts. This albino male's offspring often undergo adult molts, but not always annually like females.) Unlike some spirobolidans, *Narceus* are not sexually mature before reaching approximate adult size and post-maturity molts are commonly unaccompanied by increases in mass. The male died in February of 2007 and the female died in December 2007. The adults were almost certainly siblings and likely hatched out within a few weeks of each other. They had probably been mature for less than two years before I received them in 2004. The lifespan might be longer if provided natural cold cycles, but these specimens were never exposed to cold temperatures. The total lifespan at 70-76° F (21-24° C), appears to be about seven years for most specimens, which is short compared to *N. annularis* and *N. gordanus*.

Ten females and seven males from 2005 were retained for breeding. Eggs and 2nd instar were observed on October 2, 2007, from the second-generation adults. Some were eggs and some 2nd instars, so egg laying probably began in late August. This was much faster than I had expected. From the number of young produced (thirty-six), it is fairly certain only one of the ten adult albino females laid eggs.

Although much of their time is spent concealed in nature, the albino form is probably quickly discovered by predators. It is fantastic to see a captive-produced invertebrate variant, but what is even better is the albino *N. americanus* is extremely attractive and hardy. Unlike the regular form, the handsome color pattern is unique among millipeds. The recessive albino trait has had no negative effect on health or survival. The albinos have one of the highest survival rates for any milliped I have worked with. Hopefully hobbyists will be successful keeping and breeding this albino and it will become one of the most popular millipeds.

Northern Giant Pink
Narceus annularis (Rafinesque, 1820)
This may be the largest and most colorful of all microfauna found in the colder areas of North America, from Quebec to Tennessee. The legs, antennae and margins of each body segment are colored pink to red. The base color of the body segments is dark bluish gray. Adult females can attain a length of 4-5″, and their body is roughly the diameter of a pencil. Males are thirty percent smaller. This species was described in 1820, just three years after *N. americanus,* and remained a valid species for the next 168 years. It was changed to a subspecies by Shelley and then returned to species status by Hoffman in following taxonomic works. *Narceus annularis* seldom chemically defend themselves unless severely mistreated. However, freshly-molted individuals freely secrete quinones.

Well-fed, healthy adults breed readily in captivity twice a year. In the wild, the female lays eggs each year around mid-June, but will produce a second batch in December in captivity with reasonable conditions. None of our other *Narceus* will produce more than one brood in a year. Female *N. annularis* undergo post-ultimate molts, but they are not annual like the other two species' molts, and are also not related to egg laying.

Young millipeds begin to appear a month after the placement of breeding adults in new substrate. Eggs and hatchlings usually die if the capsule is opened, so it is difficult to determine the exact time for hatching and the first molt. The northern giant pink milliped matures slowly. Larvae are rarely an inch after a year. At room temperature this species takes at least three to five years to reach adulthood. They live at least another five years, so the total lifespan is rarely less than ten years.

Narceus annularis
Medina Co., Ohio, stock
Adults and immatures of various ages

Smokey Oak

Narceus gordanus (Chamberlin, 1943)

Large specimens grow to 0.5″ thick and are usually less than 4″ (10 cm) in length. Young animals can be confused with *Narceus americanus,* but adults are much thicker bodied and normally have yellow or orange margins, not red. The antenna groove is visibly deeper than in other *Narceus.* The head is relatively smaller, so the collum appears closer to the eyes (depending very much on the position of the head, however). Unlike *N. americanus* and *N. annularis,* mature males average the same size as females of the same stock and do not display dorsal secondary sexual characteristics. It is a common endemic in peninsular Florida.

Narceus gordanus
Small gray form

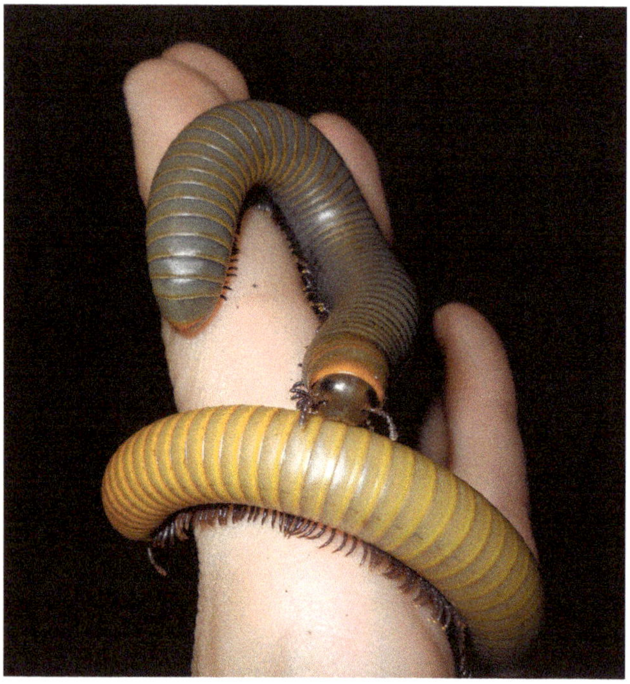

Narceus gordanus large form with orange margins and blonde form with black legs, antennae, and eyes.

Although it is found in a much smaller area than the preceding *Narceus* species, there are at least four geographic varieties that range in color and size. One common, large variety has margins with an orange cast, while the base color is slate gray with black legs. A medium-size version with orange margins is found in Orange County. There is a large blonde form from the Ocala Forest area with black legs that can have solid blonde segments or be marked with gray bands of varying width. One small variety has no orange markings and black legs, while another has thin margins with dark pinkish legs. The small gray *N. gordanus* can be found in Citrus, Hillsborough, and Sumter counties, while there's an area on Cedar Key with gray specimens that grow as large as the Ocala variety. The smaller varieties are notably smaller than *N. americanus* and *N. annularis,* but are much larger than the thin, dark, 2″ *Narceus woodruffi* that can be found in some of the same areas as *N. gordanus.*

Mature females and males molt in early summer at twelve-month intervals. About a month after molting, the adults mate and females begin to construct egg capsules. The egg capsules are visibly smaller than *N. americanus* capsules, no matter how massive the female *N. gordanus* is. Females lay

Narceus gordanus

Narceus gordanus
Large form with orange margins

Narceus gordanus
Ocala, Florida
Large form with gray and blonde banding

eggs every year, but are not terribly productive—a hundred in a brood is excellent, and greater than 60% survival past 2″ (5 cm) is good. Independent of the adult color, immatures are pale yellow until they reach a few inches in length. *Narceus gordanus* takes three to four years, sometimes longer, to reach adulthood. Wild-caught, adult specimens live seven to eight years.

Narceus woodruffi
Adults

Narceus woodruffi
Immatures

Narceus woodruffi Causey, 1959
This, the smallest and rarest of the North American *Narceus* species, is endemic to central Florida. A large adult female can reach 2″ (5 cm) in length and is bulkier than most species that size. The specimens depicted here are descended from stock collected near Gainesville in 2020. Large immatures, a molt or two from maturity, retain the pale color of tiny *Narceus* young with rows of red spots marking the ozopores down each side. This seems to be a hardy and forgiving species with little die-off if fed reasonably well. Favorite foods include rotten wood, melon rinds, and pumpkin. They breed and lay eggs at various times of year, not necessarily following a molt, but produce dozens rather than hundreds at a time.

African Giant Black and Red (AGBR)
Pelmatojulus excisus (Cook, 1897)
The African giant black and red millipeds seemed to be a possible answer to the dwindling *Archispirostreptus,* but a few enormous batches of young from the mid-2000s seem to have disappeared into the mist. Wild adult specimens from Cameroon and Ghana are still imported for European enthusiasts and are now being sold under the name 'giant fire.' Adults are massive and thick bodied like an AGB, but are relatively short and only grow to about six inches (<15 cm). An adult pair of these can go through a gallon of ground leaf mulch in a week. One female can produce more than 500 egg capsules in a season lasting up to five months. Capsules are visibly larger than the frass. The eggs are massive, about 50% bigger than the large eggs of *Archispirostreptus*. First instar larvae are also huge and have an unusually fat shape to the body. This species rarely feeds on fruit and rotten wood,

Narceus sp.
Grayson Smith / USFWS

Pelmatojulus excisus
Adult pair

Pelmatojulus ligulatus
Captive-reared adults

but consumes and requires large volumes of ground rotten leaves. It is important to keep a supply of available ground leaves for proper development of the immatures. *Pelmatojulus excisus* take five years or longer to reach adulthood.

Amber
Pelmatojulus ligulatus (Voges, 1878)
This species from Togo, Benin, and Nigeria is similar to *P. excisus* and supposedly grows as large, but even imported adults are not nearly as large and rarely exceed 4″ (10 cm). Adults are black-and-amber banded, versus the red of *P. excisus*. This is a thick-bodied species and ranks among the more massive. Adults and immatures feed almost exclusively on ground rotten leaves, but also some composts and decayed wood. Dog food and fruit are usually ignored. The ova, like those of *P. excisus*, are large compared to the capsules, but females are not nearly as productive. About two hundred is the common quantity from one female. Time to maturity is three to four years. The adult's lifespan is another three to four years with an annual molt.

Yellow-Striped Milliped
Spirobolellus immigrans (Chamberlain, 1920)
This small but striking spirobolidan has shown up only in the last few years from stock originally collected in Hawaii. It is sometimes listed on message boards as *Spirobolellus pocock,* but Pocock is the inventor of the genus name, not a valid species, and if it were described that way it would be spelled *pococki*. It is also seen listed as *Spirobolellus* sp. 'Maui' because the original specimens were collected there (Kramer, pers. comm.). Past surveys of Hawaiian animal life reported it from the island of Honolulu and called it the 'yellow striped millipede.' The Field Museum surveyed Halona Valley in Honolulu in 1997 and a specimen of '*Spirobolellus immigrans*' was identified (Evenhuis et al. 2016), though none were found in 2015 in a much smaller survey. The genus is tropical and found mostly on various Pacific islands, Australia, and Central America through the West Indies. There are at least seventy-six species of *Spirobolellus*. *S. immigrans* was the type and only species in the genus *Strophobolus* (Sierwald and Spelda 2020), now a junior synonym.

The yellow-striped milliped has uncommonly bright coloration that sets it apart from the common gray julidans that are otherwise similar in size and appearance, but there are other species with contrasting longitudinal stripes. One common, similar-looking animal is a widespread European species (*Ommatoiulus sabulosus*, the striped milliped), but *O. sabulosus* has a readily visible telson spike so it is easy to differentiate. *Spirobolellus* species do not have a telson spike and are smaller, but usually more brightly colored. The longitudinal stripes are normally very white, but they look yellow when coated in defensive secretion, which is common during handling.

The yellow-striped milliped is among the few spirobolidans that fluoresce significantly under UV light. (For most members of this order, only the tiny claws fluoresce, which would not even be visible on something this size.) The white stripes strongly glow green. *Spirobolellus* is not a bioluminescent genus, though some literature may say differently. Two *Spirobolellus* species described in Japan turned out to be misidentified *Paraspirobolus lucifugus* (Oba et al. 2011), a tiny spirobolidan that is bioluminescent, not just fluorescent.

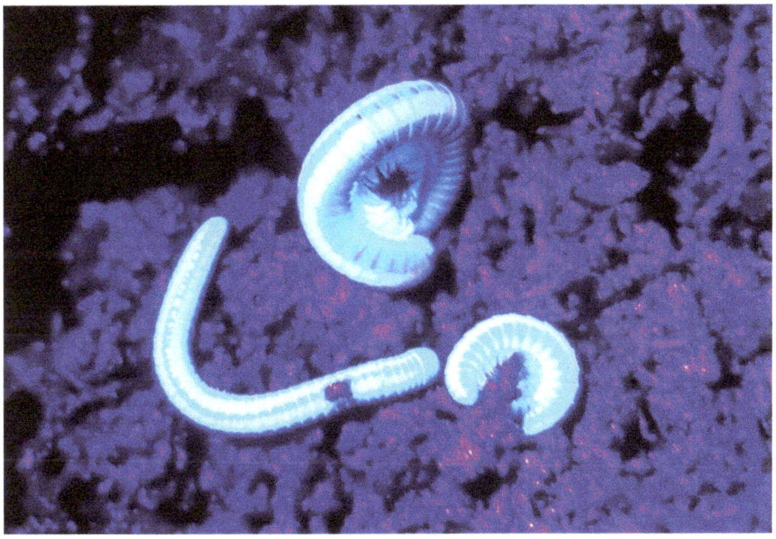

(Top and Middle)
Spirobolellus sp.
Maui, Hawaii
Under normal light and black light.

Spirobolellus sp.
Hawaii
Floyd Reed (CC BY 4.0)

Large adults are little, >1″ (up to 2.8 cm), so the enclosure should not be too large. In oversized cages the specimens can be too difficult to locate or have a difficult time locating food. *Spirobolellus* activity is primarily subterranean, so food should be buried. Specimens may not be seen for months or years if they are not dug up. A 5″ (12.7 cm) diameter plastic jar with some holes in the lid was chosen because a 3″ (7.6 cm) diameter deli-cup would probably be too small to be useful as a long-term cage, plus a deli-cup is an easy enclosure to lose track of. They feed on rotten wood along with leaves, fruit (watermelon rind and pumpkin), and kibbled dog food or fish food pellets. Again, they are small and the appetite likewise small, so the volume of food provided should be carefully considered, especially since it must be buried, or they may not eat it. You can dig them up to check on feeding and survival, but every dig risks damage, especially to molting animals.

Originally, thirteen specimens of varying size (6-25 mm) were set up with the rotten wood they were shipped in and a 1″ layer of damp/decompressed hardwood pellets was added. The supplier said the smaller ones were captive-hatched. After a few months there were large quantities of nematodes in the substrate, which could have only originated from the shipping materials because I didn't add any compost or locally collected wood. Nematodes were visible on the sides of the clear plastic. None of the millipeds were harmed by the nematodes, but nematodes eventually turn substrate into paste and can create dangerous conditions. The millipeds were carefully removed from the substrate, the container cleaned, and new substrate added. Relative to their size, they put out a good quantity and strength of quinone secretion in defense. The bright yellow liquid often coats the body, changing the white stripes to yellow. Usually something an inch or less is not as defensive, but species this size are usually julidans which tend to exude a stronger odor and little fluid. Both times they were dug up, a few specimens were curled in premolt and stayed that way until molting a few months later. Relative to size they are slower growing than any spirobolid I have kept, but still reach maturity in less than two years.

Rusty

Trigoniulus corallinus (Gervais, 1847)

This is a relatively small, drably-colored creature that was accidentally introduced to southern Florida with landscape plants, probably shipped from the Caribbean. (The ornamental plant industry also introduced the small, colorful 'bumblebee'.) *Trigoniulus corallinus* is native to Thailand and Myanmar, but has been introduced by humans to various Caribbean islands and more recently to Florida. Otherwise known as the 'scarlet' or 'rusty' milliped, it can be collected easily in great numbers in some areas of Florida. The species name means "the color of red coral" which is appropriate since red coral is dull red.

Though it had likely already been around a few years, *T. corallinus* was identified in Florida in 2006. Local collectors began to offer wild-caught Florida specimens in 2009. Due to diminutive size and unimpressive coloration, there is limited demand relative to a practically limitless supply. This species readily lay eggs and develops successfully in captivity, though it is not especially hardy. The genders are similar in size, but females average larger

Trigoniulus corallinus
Adult pair
Note the obvious hood of the male (upper right).

Trigoniulus corallinus
Broward Co., Florida
Adult

Trigoniulus corallinus
Miami-Dade Co., Florida
Adults and young
Immatures remain pale pink until the ultimate molt.

than males. Females of this species produce an egg capsule for each egg, but the structure is small, plain, and oval; very different from the unique capsules of *Trigoniulus macropygus*.

Hatchlings are white and take six months or more to gain any color. This species does not grow rapidly like the flamelegs—after two years they are only 1.5″ long and not yet fully colored. These older immatures are a delicate, light pink color. Full development, along with the change to mature coloration, takes about three years. Adults are barely 2″ (4-5 cm) and develop the trademark, rusty red (brick color) on the molt to maturity. Members of this genus are not able to breed before reaching full size and adult coloration.

A mixed diet of fruit, rotten wood, dog food, and leaf litter is recommended. Leaf litter is barely touched. Well-rotted wood seems to be the most readily accepted and significant part of the diet. The wood should be easily broken by hand and can be ground up or offered in chunks.

Philippine Flameleg
Trigoniulus macropygus Silvestri 1897
The original stock was imported in 2005 from the Negros Islands in the Philippines (just a few years before imports of tropical millipeds were discontinued in the States due to tightening of regulations). By 2023, eight generations have passed so the original stock is presently further removed from wild stock than most milliped species. This is the result of very rapid development and truncated adult lifespan relative to other giant millipeds. (If the adults lived long enough to breed with their offspring, it would be difficult or impossible to count generations.)

The adults are a very colorful creature whose long legs are red at the base but transform to bright yellow, evoking the common name. There is red banding on the body segments, but this coloration is muted (sometimes exaggerated by lighting in photos), and has nothing to do with the common name. Red bands between the segments develop only after the ultimate molt and some keepers have confused the solid black of late instar specimens with a different form or species.

In addition to spectacular daylight coloration, this species fluoresces under black light. The body glows a muted dark green like some other giant millipeds. In stark contrast to nearly all fluorescing terrestrial invertebrates that glow green, the legs of this species glow bright orange, adding another dimension to the name Flameleg. The species name *macropygus* means "large rump," is minimally descriptive, and probably refers to the subjectively large and elongate paraprocts.

Adult females are often half again more massive than the males and grow to 4-5″ (10-12 cm), males rarely reach 4″ (up to ~9 cm). Adults are easy to differentiate from a distance by size and the slightly hooded appearance of the male. In captivity females tend to wait six to twelve months after reaching maturity to lay eggs, although they can mate within weeks of the ultimate molt. A female constructs capsules over a period of three months to a year and produces up to 400 hundred eggs. Each is laid inside an unusual egg capsule that could be called an alien ice cream cone. The rounded 'ice cream' top has a dozen large dimples, while the cone is a malformed piece of frass. This is the most unusual egg capsule presently known. This species does not undergo post-ultimate molts, which is unusual for a

Trigoniulus macropygus
Negros Island, Philippines
Adult male upper right
Subadult female lower left

Trigoniulus macropygus
Negros Island, Philippines
Specimens have red margins at maturity, but not as subadults. This has lead to some confusion whether there are two different color forms or species.

Trigoniulus macropygus
Fifth generation (2012)

spirobolidan. Females die shortly after egg laying has ended. A second reproductive season would be a rare exception. Males live one to two years after maturity and many outlive females.

Trigoniulus macropygus feed on a variety of foods commonly consumed by millipeds but should be supplied with, and can do very poorly without, ample, shredded, rotten leaf litter. With reasonable feeding, the time from egg to adult is fourteen to sixteen months.

California Red Stripe
Tylobolus claremontus Chamberlin, 1918
This a readily available and pretty species that is 3-3.5″ (7-8 cm) long at adulthood. It feeds on rotten wood and leaves, as well as dog food and various fruits. Females produce normal spirobolidan egg capsules, though young may not be easy to rear. Adults live at least a few years.

Tylobolus uncigerus (Wood, 1864)
Although *T. uncigerus* is common from the San Francisco Bay northward into southern Washington, I had only seen a dozen specimens and only maintained one in captivity long term before 2017. That one from 2008 was an adult that lived a year and a half. It was extremely defensive and prone to release defensive exudate at the slightest provocation. However, the current specimens and offspring hardly ever exude fluids with normal handling.

Mature *T. uncigerus* are usually a steel gray color, rarely brown. Interesting highlights are pale rings between each leg segment and a very thin margin on the body segments. Under 365 nanometer UV there is a pale but noticeable green fluorescence, mostly on the pale margins. Specimens are huge (7-8 cm) compared to bumblebees (*Anadenobolus monilicornis*), but are barely the size of a young *Narceus americanus*.

In 2017 I spoke to a hobbyist who collected some local *T. uncigerus* in Washington state and produced fewer than a hundred tiny immatures. I am usually excited about captive reproduction because it shows it is possible to breed the given species in captivity, beyond simply knowing the specimens did not originate in the wild. With captive specimens I expect the best practical health and a certain ease of care in captivity. For a trade, I received an adult female and a number of immatures that arrived 11-17-2017. The latter were about 0.5″ long (12mm) and were probably laid as eggs in June or July of the same year.

Over the following years the young grew at a slow but consistent rate. The only deaths noted were a few specimens interrupted during molting. They were accidentally damaged when the substrate was replenished or when specimens were checked on. I traded quite a few away and was not sure they had reached maturity. The female had continued to molt once a year every February or March and by 2021 was still healthy.

The larger young were approaching mature size by late 2020. Diplopods, depending on the species, can usually reproduce at half to two-thirds full size and continue to grow after initial reproduction. I did not observe mating, but in February 2021 I dug through the substrate and discovered a damaged egg capsule containing a healthy-looking egg. Healthy eggs are pale yellow to white and fully inflated. I dug through and found another. I stopped excavating because eggs in damaged capsules often die.

Tylobolus uncigerus probably do not lay eggs during the winter months in nature, but

Tylobolus claremontus
California

Tylobolus uncigerus
Female with young (2017)

Tylobolus uncigerus
Same female with different offspring (2021)

these were kept around room temperature. The enclosure rarely gets cooler than 70° F (21° C) even in mid-winter. The eggs were probably laid by the original female that mated with one of the young males (these may or may not have been her offspring in 2017), because the other females were still only about seventy percent her size.

By the end of 2021 the next-generation specimens averaged 1.2″ (3 cm), suggesting they were about four months ahead of the seasonal development outdoors. I counted sixty-two healthy specimens. This number did not include eight lost to handling damage since spring and three molting cell deaths due to limited oxygen. This occurred when the substrate became very damp, and cells at the bottom of the substrate did not get enough oxygen. Limited ventilation and use of melons rinds can lead to this problem.

The adult female (and some of the mature offspring) are readying to molt again, probably in February 2023. However, the offspring have not quite reached her size. Time taken to reach maturity for this species is three and half to five years and adult longevity can exceed five years. I anticipate the original female will lay eggs again in late spring. I am surprised by her longevity only because she has often been provided sub-optimal conditions. There were problems with soil nematodes and more recently sub-starvation caused by a switch to wood particle substrate (expanded hardwood pellets).

Tylobolus utahensis (Chamberlin, 1925)
This is another *Tylobolus* that is very common in the wild, but I have never seen it on a vendor list. It is found in Utah, Nevada, and Arizona. I acquired one specimen from southern Arizona 2007 that was an adult male. It lived for three years and underwent two post-ultimate molts, but did not grow longer than 3″ (7.6 cm).

Spirobolus bungii; Xinyang, Henan, China
Zinogre (CC BY-SA 4.0)

Tylobolus uncigerus
These can be reddish or gray at maturity.

Tylobolus sp.
Marin Co., California
L. J. Moore-McClelland (CC BY 4.0)

Trachelomegalus sp.
Sarawak, Malaysia
Bernard Dupont (CC BY-SA 2.0)

(Order Spirobolida) *Acanthiulus* sp.; Papua New Guinea
Coenobita (CC BY 4.0)

(Order Spirobolida) *Eucentrobolus* sp.; India
T. R. Shankar Raman (CC BY 4.0)

(Order Spirobolida) *Messicobolus magnificus;* Chiapas, Mexico
Daniel Pineda Vera (CC BY 4.0)

(Order Spirobolida) Red-Spined Milliped, *Xenobolus carnifex;* Chennai, Tamil Nadu, India
Paulmathi Vinod (CC BY 4.0)

Antheromorpha rosea
Thailand
Male

Antheromorpha uncinata
Thailand
Female

Apheloria montana
Cosby, Tennessee
MBWildlife (CC BY 4.0)

Order Polydesmida (Flat Millipeds)
This is the largest order with 3,500 species. All species are known to produce hydrogen cyanide in defense. *Poly* means 'many' and *desmos* means the 'links of a chain.'

Antheromorpha rosea Golovatch 2013
This genus contains some of the largest members of the Family Paradoxosomatidae. Most are small like the well-known and ubiquitous greenhouse milliped, *Oxidus gracilus,* a species commonly confused with a handful of similar-looking, related paradoxosomatids found in garden beds and greenhouses, also introduced from Asia. The bright colors, unique markings, and large size makes *A. rosea* difficult to confuse with other members of the family. Adults mature at up to an 1.25" (ranging 34-44 mm), with females on the longer end of the spectrum. Imports from Thailand showed up in captive stocks in 2022. They seem hardy but are probably short-lived. Likhitrakarn et al. (2016) report that (in Thailand at least) adults are only seen for one or two weeks once a year, in September or October. In captivity they live at least through the next spring. Living specimens from Thailand are usually orange, while Chinese specimens are red. However, the species name refers to the pale pink of preserved specimens from Yunnan Province, China.

Antheromorpha uncinata (Attems, 1931)
Adults of this warningly-colored species are marked in dark red, orange, or pale yellow with various intergrades. The color pattern is easily discerned from *A. rosea* even if individual specimen tint overlaps—though it is similar in size, it is brighter in color. The contrasting body markings and legs range from pale gray to dark black. It is not possible to tell what color the adults will be before the ultimate molt since the immatures are pale and vary little in coloration. This large paradoxosomatid grows to between 30 mm and 44 mm at maturity. Males are visibly smaller than females. This species is currently only known from Thailand. It has been observed displaying swarming behavior at Pang Rimkorn, Chiang Rai Province (Likhitrakarn et al. 2016). While these showed up only recently, there is a prior photo of a live specimen given as an example of foot-rot (Sigling 2010 p. 59, fig. 98).

Apheloria virginiensis (Drury, 1770)
Family Xystodesmidae is an immensely varied group in North America that includes the genera *Apheloria*, *Cherokia*, *Dicellarius*, *Furcillaria*, *Nannaria*, *Pachydesmus*, *Pleuroloma*, and *Harpaphe,* among others. Many have geographic and subspecies color patterns which resemble patterns of other species, confusing identification. External identification of similar-looking species and subspecies (without gonopod dissection) relates to the degree of arching or vaulting of the body, tergite texture, color pattern, etc., that can be rather subtle (Hoffman, pers. comm., 2005). Adults range in length from 1-3.1" (2.5-8 cm) by species or subspecies. The genus *Apheloria* contains four species, four subspecies, and numerous color forms, with a distribution from southeastern Canada south to Georgia, and west to Arkansas. Adults are 2" or more (5-6 cm).

Apheloria virginiensis is among the most beautiful and striking of the large North American flat millipeds. They also fluoresce brightly under UV. Mature specimens are marked with a contrasting black-and-yellow

Apheloria virginiensis corrugata
Washington, Virginia
Judy Gallagher (CC BY 2.0)

Apheloria virginiensis
Orange Co., New York
Kristof Zyskowski (CC BY 4.0)

Boraria deturkiana
Haywood Co., North Carolina
Evan M. Raskin (CC BY 4.0)

Apheloria virginiensis
Hocking Co., Ohio
Pale subadult at top right

Apheloria virginiensis
Hocking Co., Ohio

Boraria stricta
Cane River, North Carolina
Jack Byrley (CC BY 4.0)

Cherokia georgiana
Canton, Georgia
Christina Butler (CC BY 2.0)

Coromos vittatus vittatus
Adult
© Tom Larsen

Desmoxytes purpurosea
(A) Male, (B) Female
Srisonchai, R., H. Enghoff, N. Likhitrakarn, S. Panha. 2018. A revision of dragon millipedes I: genus *Desmoxytes* Chamberlin, 1923, with the description of eight new species (Diplopoda, Polydesmida, Paradoxosomatidae). *ZooKeys* 761: 1-177. (CC BY 4.0)

pattern. From some locations, the yellow fades to red or pink towards the paranota. A small percent of specimens from Hocking County, Ohio, have red stripes instead of yellow. Immature specimens are pale white until they reach the subadult stage, which is a pale version of the adult pattern. It is possible to raise immatures and keep adults alive a year or so, but a few months is more common. Warm temperatures and major disturbances kill them. *Apheloria* are usually found in small depressions beneath rotten logs on sloping or steep ground. They seldom congregate in nature, but there can be a few in different depressions under a large log. They eat rotten wood, but also consume sandy clay soil, which improves long-term success under laboratory conditions.

Nigerian Flat
Coromus vittatus vittatus (Cook, 1896)
This huge, central African (Cameroon, Liberia, and Nigeria) species can grow larger than the biggest North American *Pachydesmus* and reaches around 3.5″ (9 cm). I do not remember ever seeing specimens during the United States pre-ban era, but it always seemed to be a popular import in Europe. The imported specimens (when collection location was provided) were from Nigeria. The adults seldom hide and therefore the defensive fluid is copious. This is one of the few species that might be able to poison itself, and other inhabitants of the enclosure, if ventilation is not adequate. Decomposing leaves and wood from beech and oak have been used as food. They are very gregarious and live in colonies, even at first instar. Eggs have been laid and hatched a few times in captivity—the adult females tend to die shortly afterwards. Larvae are tiny, white, and hatch out in mass at the same size, so eggs are probably laid in a large group in a single chamber like *Pachydesmus*. Attempts at keeping hatchlings from eggs laid by wild-caught females alive, past the first few molts, have not been successful.

Pink Dragon
Desmoxytes planata Pocock, 1895
Most members of this genus are restricted to small areas, but *D. planata,* once found only in the Andaman Islands, has been spread by humans across Southeast Asia including Fiji, Java, Sri Lanka, Thailand, and the Malay peninsula. They are beautiful creatures, but they are small and reports of specimens longer than 1.2″ (3 cm) lack substance. The largest adult depicted here is 0.9″ (2.2 cm). The pictures included depict captive-bred specimens and like many of the small paradoxosomatids, they have proven relatively easy to reproduce. Adults seldom live six months, so if eggs are not laid the culture comes to a rapid end. *Desmoxytes* are members of the Family Paradoxosomatidae, one of the largest orders with around 1,000 species. It is a big family of small species, and most are less than an inch in length. *Planata* means 'flat' and *desmoxytes* can translate as 'a long portico of links.' This genus first became widely familiar when the 'shocking pink dragon' *Desmoxytes purpurosea* was placed on the top ten list of new species in 2008. The shocking pink dragon looks much like the pink dragon, but it is limited to limestone caverns in Uthai Thani Province, Thailand.

Leach's Colored Train
Euryurus leachii leachii (Gray, 1832)
This subspecies from Illinois, Indiana, Kentucky, and Ohio is the namesake for a small

Desmoxytes planata

Euryurus leachii
Ashland Co., Ohio
Adults and immatures

Euryurus maculatus
Liberty Co., Florida
The black earthen ring is frass.

Euryurus sp.
Muskingum Co., Ohio

tribe of North American xystodesmids, the Euryurini. One other subspecies, *E. leachii fraternus* (Hoffman 1978), is found from western Pennsylvania to northern Alabama. The body is dark bluish gray with hot pink spots on the sides and middle of each segment. The hot pink spots often appear orange in photographs. The legs are white.

A dozen adult specimens can be kept in a six quart (5.7 L) plastic shoebox provided with an inch of well-rotted wood. (No other substrate materials should be used.) Rotten wood should be collected from a hardwood forest, but at a high level of decay so it is difficult to determine what species of trees have been used. Wood should be dark brown and easily crumbled when wet, not black or mushy. Old enclosures begin to look like crazy cities built out of frass. As soon as the substrate has been mostly consumed, to prolong the use of the culture container, 10-25% cottonwood sawdust can be mixed in. Within a few months, the sawdust starts to be broken down by fungi and partly consumed by the millipeds. A piece of moist and meaty, steak-flavored dog food attracts feeding, but it should only be a tiny portion of the diet. Any uneaten portion should be removed after a few days. If there are later instar offspring or if specimens greatly reduce frass production, a new enclosure with fresh wood must be set up. If the keeper waits until specimens start dying, the rest may be unable to recover from starvation.

After adults are observed mating, it will seem like frass production has kicked into overdrive. Hundreds of tiny white eggs will be pasted inside small capsules formed by rectal secretions. Each is deposited in its own chamber resembling a tiny mud pie. Frass capsules are cemented everywhere along the sides of the cage and against pieces of soft, dark wood. Forming hundreds of individual capsules takes the female a few months—larvae hatch over the same time length. The first half will reach maturity in less than a year, while the second half can take a few months longer. If not, they slow down and require an additional year. Adults usually live six months, rarely more than a year and a half.

Spotted Train
Euryurus maculatus (Koch, 1847)
This species is very similar to *E. leachii* and has also bred readily in captivity in adequate habitat of well-decomposed wood—eggs are likewise deposited singly. The color is slightly different, but male gonopod dissection would be required to differentiate the species if the collection location were unknown. This *Euryurus* has a southern distribution: Alabama, northern Florida, and Georgia. Eggs are also laid separately in cells formed in mounds of pasted frass. It seems to have a much better survival rate (based on the captive development of a single large group of each species, each produced by a small number of females from a single collection locale). Immatures take one to two years to reach adulthood.

Harpaphe haydeniana (Wood, 1864)
There are six different subspecies along the west coast of North America, from Alaska to Canada, south through northern California. This is a pretty and large xystodesmid that survives about a month (Washington specimens) or a year (northern California ssp.) in captivity at room temperature. Sensitivity to shipping and temperature seems to depend on collection location. The longer-lived ones might breed easily in a proper environment (a few inches of

Harpaphe haydeniana
Washington

Harpaphe haydeniana
Oregon
Peter Pearsall / USFWS

Helicorthomorpha holstii
Florida

crushed, brown, rotten wood that has been heat-treated for pests and kept very damp). Specimens can be collected in huge numbers using black lights in mid-summer.

Orange Rosary Milliped
Helicorthomorpha holstii (Pocock, 1895)
This species was originally described from a type specimen in Okinawa, Japan. It is widespread across southeast and East Asia and southern China and is very common in Vietnam (Golovatch 2020). More recently this species was introduced to Florida (Shelley and Edwards 2001). Members of the Family Paradoxosomatidae are easy to discern from other flat millipeds by the large groove across the top of each pseudosegment. *Helicorthomorpha holstii* is easy to differentiate from other family members by its unique, colorful markings.

Collectors in Florida started intermittently offering wild-collected specimens of this species a few years ago. The adults are quite handsome in close-up images, but in person they are not as spectacular because small size makes it difficult to see the pattern of colors. Adults are barely longer than two centimeters. Features that are obvious in a photo are more difficult to see in actual size.

A loose-fitting, but solid, glass lid allows for adequate ventilation while maintaining humidity well. Dry conditions can result in rapid death. Adult *H. holstii* have consumed notable quantities of hair and slime forms of freshwater algae (harvested from axolotl and Spanish newt aquariums). This species probably feeds on common milliped fares including fish food, fruit, and rotten leaves, but not in measurable quantity. Some keepers have reported egg laying and development, but hatchlings should be observed within a month or two. Adult specimens rarely live longer than five months.

Ohio Twisted-Claw
Nannaria ohionis Loomis and Hoffman, 1948
Members of *Nannaria* are known as twisted claw millipeds, and with seventy-eight described species it is the largest xystodesmid genus. *Nannaria ohionis* is endemic to Ohio and adults are around 1.2″ (3 cm). Adults are usually pale (rarely dark) and resemble an immature *Apheloria*. They are substrate feeders, and it is unusual to see them nibble on any type of supplemental food. Powdered rotten wood and leaves can be mixed into clay soil for feeding. Temperatures no higher than 75° F (24° C) are suggested.

Long-Flange
Orthomorpha coarctata (Saussure, 1860)
This Asian paradoxosomatid has been spread by man across the tropics and subtropics of Earth (though the first generation of the specimens depicted here came from Oga's Cantina on Batuu). This species is fascinating because mature specimens have a popular xystodesmid warning coloration: a black body and yellow spotted paranota. They look like a tiny version of similarly colored *Harpaphe* and *Apheloria*. Adults are 15-27 mm with males on the short end of the size range. Like many members of its family, these are easy to breed and rear in captive enclosures. They breed and develop well at 75° F (24° C). Food consists of decayed leaves and vegetable detritus, but also some algae and melon. Young develop through the second stadia within four weeks of egg laying and mature in five months. Late-stadia specimens are usually dark gray but can be white. Adults live six to twelve months after the ultimate molt.

Motyxia monica
Tehachapi, California
Ken-ichi Ueda (CC 0)

Nannaria ohionis
Athens Co., Ohio

Orthomorpha subkarschi
Thailand
Adult male

Orthomorpha subkarschi
Adult female in brooding chamber (with the top cut off)
The brood chamber is made days or weeks after mating and is similar to a molting chamber constructed of many frass particles.

Orthomorpha subkarschi
These immatures are well developed upon leaving the brood chamber, as big as a subadult *O. gracilis*.

Orthomorpha subkarschi
Late stadia immatures

Yellow Dragon
Orthomorpha subkarschi Golovatch, 1998
This significantly large species of *Orthomorpha* (adult males around 4.5 cm, females 5.5 cm) has also been seen under the name 'yellow banded.' Specimens made their way into collections wild caught from Thailand. The adults are substrate feeders and show little interest in leaves, fruit, or dog food. A group of six was maintained at 74° F (23° C) for four months, were moved to a new enclosure at 79° F (26° C), and died one by one over the next four weeks. Females cement themselves inside a large frass-lined cavern in the soil to lay eggs. Immatures pass a number of molts within the chamber, surfacing around the time the mother dies. At this age they gorge on dog food. They also feed on substrate, decayed wood, and rotten leaves.

Orthomorpha coarctata
Broward Co., Florida

Pachydesmus crassicutis (Wood, 1864)
The largest xystodesmid in North America, north of Mexico, resembles various large polydesmidans from the Old World. Adults are plus or minus 3″ (7 to 8 cm) in length and occur from central Mississippi through eastern and south central Louisiana (with specimens depicted from Lafayette). There had been eight subspecies described between 1899 and 1958, but they were elevated to species in 2021. The former subspecies range across the southeast: Alabama, North and South Carolina, Florida, Georgia, Louisiana, Mississippi, and Texas. Adults are long and thick bodied, but not as wide as a large *Apheloria* because the expanded plates (paranota) on the sides are shorter. Overall color is tan to brown with pale, ivory-colored paranota.

Adults commonly survive one to two years in captivity. Wild-caught females lay eggs around October in captivity and do not require recent mating with a male as an egg-laying stimulus. Each female creates a single, large, hollow cell, cemented together with hundreds of frass particles. Inside a few hundred eggs are laid in a large, sticky clump. The eggs in each giant capsule hatch the same day. The early instars have proven difficult to keep alive. Hatchlings dry out rapidly if there are any dry areas of the substrate. Conversely, they seek out condensation and will drown in water droplets. They feed on well-rotted wood primarily, but also fish pellets and dog food.

Platyrhacus sp. 'South Thailand'
These large, slow-moving specimens are sometimes listed under the name *P. marginellus,* a similar species described from Sumatra, Indonesia. There are around forty species in the genus. The Family Platyrhacidae

Pachydesmus crassicutis crassicutis
Lafayette Parish, Louisiana

South Carolina polydesmids from clay cells beneath a large log
Daylight (middle)
Ultraviolet light (bottom)

Platyrhacus sp.
Southern Thailand

includes many bulky species across the world's tropics, including the South American *Nyssodesmus* that reach up to 4″ (10 cm). Adults of 'South Thailand' are plus or minus 3″ (6.5 to 8 cm) in length. Like xystodesmids, the living specimens produce piles of frass structures covering the enclosure substrate entirely over time. They feed on detritus and soil and have little interest in supplemental foods. Anecdotal shipping mishaps suggest this species is extremely sensitive to cold temperatures. They become motionless around 50° F (10° C) and die within hours from cold temperatures well above freezing.

Pseudopolydesmus canadensis (Newport, 1814)
This is a common species found individually beneath rotten logs and bark in wooded areas. It ranges across the Appalachian Mountains and foothills of eastern Canada and northeastern U.S. The semi-hardy specimens feed on a variety of foods, mostly decaying leaves and some fruits. Like other members of the Family Polydesmida, specimens do not fluoresce brightly green under ultraviolet, but the pale legs glow blue under 395 nm. Females build a large frass cell under wood or rocks, lay a group of eggs, and curl around them. Only one batch can be produced because she dies around the time they hatch. Specimens live one to two years after hatching and die shortly after reproduction.

Sigmoria spp.
Specimens of a pink and black *Sigmoria* sp. from Mississippi were available seasonally in the 1990s. The wild-caught adults survived a few months, but almost never lived longer. The reason they survive such a short time in captivity is still uncertain, but many of the adult xystodesmids are seasonal and none undergo post-ultimate molts. Fifteen *Sigmoria* species and seven subspecies are known to occur in the southeastern United States (Hoffman 1999). Adults are usually

Platyrhacus face

Pseudopolydemus canadensis
Central Pennsylvania

Pseudopolydesmus erasus
Canton, Georgia
Christina Butler (CC BY 2.0)

Pseudopolydesmus serratus
Northwestern Pennsylvania
1.2" in length
Easy to keep alive for a year or so.

Pseudopolydesmus serratus
Washington, D.C.
Katja Schulz (CC BY 2.0)

Semionellus placidus (?)
Sumerduck, Virginia
Judy Gallagher (CC BY 2.0)

Sigmoria trimaculata
Male gonopods

Sigmoria trimaculata
Sparta, Tennessee
Malcolm Manners (CC BY 2.0)

Wandering Cherry
Pleuroloma flavipes
Indiana
David Beamer (CC BY-SA 4.0)

Xystocheir dissecta
Don Loarie (CC BY 2.0)

found under leaf litter, which is odd for a xystodesmid. Most species shelter in small scrapes beneath logs. When collecting, keep in mind they can only survive the heat of a closed hand for a very short time. *Sigmoria* spp. resemble *Apheloria* but are often smaller—the coloration is red or pink with black, and is not nearly as brilliant.

Sigmoria sp.
Sevier Co., Tennessee

Order Sphaerotheriida (Giant Pill Millipeds)
Members of the Order Sphaerotheriida are known as giant pill millipeds and had been popular U.S. imports, mostly from Madagascar, from the early 1970s through 2006. They have twelve visible body segments and forty-two legs and are easily differentiated from the 'small' pill millipedes in the Order Glomerida. Massive specimens from Madagascar, as big as a small orange, and marble-sized Indian adventives to Tanzania had been seasonally available on North American invertebrate vendor lists for decades. Other species from Southeast Asia are rarely seen.

Tanzanian Pill
Arthrosphaera brandtii (Humbert, 1865)
Specimens had been imported to the North American and European markets for a number of years, but different batches were always the same straw tan color (unlike the variably colored Madagascans). Five or ten percent of specimens often died in the initial weeks from shipping stress, but any survivors would live two years within adequate artificial habitats. They feed on substrate and pay little attention to fruit or dog food. Imports of this species continue in Europe to the present day. One keeper documented collection of eggs from a wild female, and early development of immatures in 2007, but there have been no repeat performances. These grow to the size of a large marble; uncurled length is around 1.25" (~3 cm). Males tend to be smaller than females, are much less common, and have a faint to obvious circular dent in the middle of the anal shield. This species is not native to Africa. They were accidentally introduced to Amani, Tanzania, from India by an agriculture station. The natural range of *A. brandtii* is India and Sri Lanka.

Castanotherium sp. ex. Philippines

Unlike *A. brandtii* above and *Zoosphaerium* below, which were imported by the tens of thousands over a few decades, I am only aware of two specimens of this pill milliped making it to the United States (though many were seen in European collections in the early 2000s). Adult size is similar to *Arthrosphaera,* around 1.25″ unrolled (~3 cm). Specimens are dark brown with yellow to red highlights near the segment margins. The U.S. imports lived about thirty months and did not reproduce; both were females.

Thai Giant Pill

Zephronia enghoffi (Srisonchai, Sutcharit, Likhitrakam, 2021)

This is one of the larger species from Thailand, with adults growing to 1.5″ in length (29-36 mm). Specimens are from Khon Kaen and Loei Provinces. They are active and heavy feeders on substrate, well-decayed wood and leaves, but also feed on melon rinds. Still, they do not live long. Like other giant pills, it can take many minutes to hours for a specimen to uncurl, but the gender can be determined externally. The anal shield of the female is rounded while the male's is rectangular (Srisonchai et al. 2021).

Green and Bronze Pill

Zephronia siamensis Hirst, 1907

This beautiful sphaerotheriidan is endemic to eastern Thailand and was the first member of the genus described from that country. The coloration of specimens in the attached images (copper square markings) is relegated to Srichang Island specimens. Adults range in size from 15-25 mm. The male's anal shield is marked differently, having a shallow dent lengthwise across the middle.

Giant Green Pill

Zoosphaerium sp. ex. Madagascar

Large specimens are arguably the most awe-inspiring millipeds in the world as they are almost as wide as long. Specimen length ranges from 1.5-3.5″ (4-9 cm). Imported creatures often come in a spectacular, glowing forest green, but some groups have come in brown, orange, or green with red margins. The different colors and sizes may represent different species.

All species/varieties are notoriously poor shippers and live a few days to a month after purchase. Unsupported rumors suggested lichens, algae, mushrooms, and other foods could keep these pill millipeds alive. (*Zoosphaerium* do not eat freshwater or saltwater algae and have almost no interest in lichens. Unrelated terrestrial isopods relish algae and lichens, which might have led to the rumor.) Healthier specimens will turn large quantities of substrate (compost, rotten wood, and leaves) into frass pellets. Food may have had little effect; the majority

Zephronia enghoffi
Male claspers

Castanotherium pill millipeds
Philippines

Zoosphaerium sp.
Madagascar

Zoosphaerium sp.
Madagascan pill in defensive sphere

Zephronia enghoffi; Phu Laen Kha National Park, Thailand
Len Worthington (CC BY-SA 2.0)

Zephroniidae; Sarawak, Malaysia
Pavel Kirillov (CC BY-SA 2.0)

Zoosphaerium sp.; Falierana, Andasibe, Alaotra Mangoro, Madagascar
Pavel Kirillov (CC BY-SA 2.0)

Zoosphaerium sp.; Falierana, Andasibe, Alaotra Mangoro, Madagascar
Pavel Kirillov (CC BY-SA 2.0)

Zephronia enghoffi

Zephronia enghoffi

Zephronia siamensis
Gender chromatic dimorphism of the anal shield
Female (left), male (right)

Zephronia siamensis

Glomeris klugii

Glomeris klugii

of specimens died within a few weeks and did not eat at all. It seems they were already effectively dead, and it is unlikely they could have been saved by any food or habitat parameter in such a short time. Although *Zoosphaerium* likely live more than a decade in the wild, average imports survive a month. The maximum verifiable time in an artificial habitat is seven months, attributed to the specimen in the molt photograph from 1995. No keepers have presented photographs of eggs, hatchlings, or other stages. This molt sequence is the first photographic documentation for any part of the regular life cycle in an artificial habitat. *Zoosphaerium* seem especially sensitive to temperatures outside the 65-85° F (18-27° C) range, but they may just stress easily.

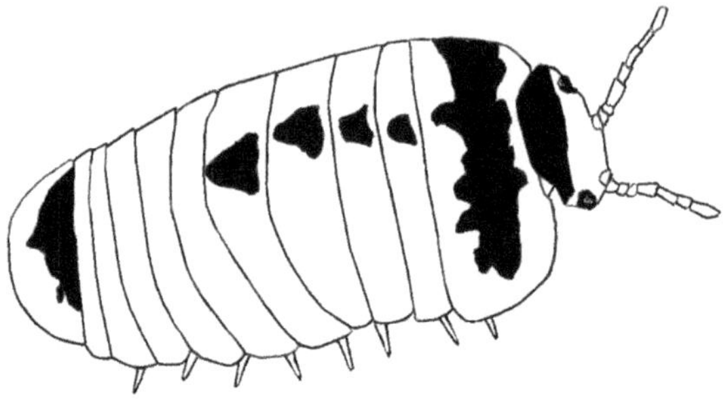

Order Glomerida

Glomerida is a small order of millipeds with less than five hundred species. While representative species can be found on all continents except Antarctica, the group has only been successful in warm temperate areas of the Old World. The largest genus in the Glomerida is *Glomeris*—it contains about 100 species, and a few hundred subspecies, mostly from Europe (Golovatch et al. 2009). *Glomeris* comes from the Latin word for 'ball.'

The order superficially resembles the large pill millipeds (Order Sphaerotheriida) that were popular when U.S. imports were allowed (2006 and prior). Despite the similar appearance, Glomerida is a unique order with fewer visible body segments, fewer legs, and a unique chemical defense system. Glomeridans have twelve body segments, but appear to have eleven. Males have thirty-four legs and females have thirty-eight. The massive, thick-shelled Sphaerotheriida have no chemical defense, while the Glomerida produce powerful substances related to Quaalude (Eisner 2003). For zoologists, the most important difference is that Sphaerotheriida do poorly under laboratory conditions, whereas Glomerida can survive, undergo regular post-ultimate molts, and reproduce. At least one species had been reared from egg to adult under laboratory conditions (Iatrou and Stamou 1990) and another species has been documented through consecutive generations (McMonigle 2020). *Onomeris* and *Rhopalomeris* have also laid eggs, with some success in development of immatures.

Harlequin Pill
Glomeris klugii Brandt, 1833
This cryptic beauty may be found throughout Europe and in parts of northern Africa,

but the primary range is eastern France, northern Italy, Austria, Germany, Slovenia, Croatia, and the Czech Republic.

Other *Glomeris* maintained in captivity, like *G. pustulatus,* do not seem to produce a noticeable defensive secretion during careful handling at any time. *Glomeris klugii* do not always produce a defensive secretion, but they often produce notable secretions with very limited handling (less than 30 seconds). All specimens in the accompanying photographs have produced defensive fluids, but not all the time. The defensive secretion is a clear, sticky fluid exuded from the middle of the dorsal surface between the segments. It occurs when a specimen is handled and already in the shape of a ball. The fluid does not feel sticky, but when you touch two fingers together and pull them apart you can see it congeal. There is no odor associated with the defensive secretion. Also, it does not cause skin discoloration or discomfort (McMonigle 2018).

The largest of four females was 8 mm wide and 16.5 mm long, measured while walking (the next largest specimen being 16 mm). Maximum diameter is 9 mm conglobated. It only takes a few minutes to measure a specimen walking because they do not always roll into a ball when disturbed. The six males were smaller—13-14 mm in length when naturally extended. Other than smaller size, there are no obvious differences in the body armor of the male and female. However, the male's rear legs are adapted into grasping organs called telopods.

Pale, dark, and reddish color forms can be collected from the same area. All three primary color variations were seen in eight specimens from one collection locale. Color is variable for both genders, but males tend to be lighter.

Substrate provided was composted manure mixed with expanded hardwood pellets. Shredded, decaying hardwood leaves were added on one side and a piece of bark with crustose and foliose lichen on the other side. The primary food (as evidenced by piles of black frass) was the compost. They also fed on the fresh crustose lichen, but not the leafy type. Rotten wood and a mushroom collected outside did not show any evidence of feeding. A piece of kibbled dog food was nibbled. With the above diet, they seemed very healthy for six months but grew weaker over time and some died during the frequent molts (two or three times a year as adults) until all were dead by twenty months.

Glomeris pulchra Koch C.L., 1847

This beautiful European species is found throughout the Balkans, including Albania, Croatia, Hungary, Macedonia, and Slovenia. The species name *pulchra* means 'beautiful.' More than two dozen varieties have been proposed for different patterns in the spots and colors across geographic regions, but they are taxonomically invalid. The body is brown to black, while the spots are yellow, orange, red, or a combination. This species somewhat resembles *G. pustulata* with fewer spots, but it grows larger and matures more slowly. Offspring grow quickly in the first year or two, but then molt just once or twice a year before reaching mature size around four years. This species has been reared through a second generation under artificial conditions.

Glomeris pustulata (Fabricius, 1781)

The type species for the genus is *G. pustulata. Pustulata* means 'blister' and probably refers to the defensive secretion—it shares this species name with some plants

Glomeris pulchra

Glomeris pulchra
Immatures

Glomeris pustulata
Immatures

that have oil glands. This species is found across warmer areas of Europe and a few subspecies have been described based on geographic coloration.

The colorful spots on *G. pustulata* advertise its defensive poison. The warning coloration of another familiar *Glomeris* species, *G. marginata,* is mimicked by the zebra pill bug (*Armadillidium maculatum*). This seems to be a clear-cut case of Batesian mimicry, since the isopods do not have chemical defenses. The red spots on some *Armadillidium* are thought to mimic the Mediterranean black widow *Latrodectus tredecimguttatus.* (The spider may drop to the ground when disturbed, and the brightly-colored spotted abdomen is similar in shape and size to a rolled-up pill bug.) This is another case of Batesian Mimicry where the isopods benefit from a well-protected model. However, there are spotted *Glomeris* in the same areas of southern Europe as the *Armadillidium* and *Latrodectus,* so the Batesian mimicry of the *Armadillidium* could benefit from additional models provided by the Muellerian mimicry of *Glomeris* and *Latrodectus.* The habitat preference of *Glomeris* and greater population density (predatory *Latrodectus* are fewer in number and not found in aggregations) suggest these pill millipeds are the more likely model for the *Armadillidium* mimics. This interaction may be part of the reason southern Europe is the hotbed for both isopods and pill milliped species.

Each animal molts more than a dozen times on its path to maturity. Like sphaerotheriidans, they do not appear to eat their shed exoskeletons. Unlike the very difficult to keep alive sphaerotheriidans, glomeridans molt regularly in captivity. Like flat millipedes (Order Polydesmida), they construct a thick-walled molting cell fashioned out of pasted materials, but there are differences. First, the cell is buried in the soil rather than attached to wood. Second, the outside surface is without shape rather than covered in small lumps (and the shape of individual frass pellets is not easily seen). Finally, there is no neck or cap, as it has a rough spherical shape.

Food preference in captivity is difficult to gauge. Light brown rotten wood and leaves do not appear to show up in the pellets, which are mostly black like the composted manure substrate. Darker brown rotted-wood pellets have been observed. Small pieces of watermelon, cantaloupe, and (freshwater) algae are eaten, but are not discernable in the frass. Animals are usually found on these foods a day or two after the food is placed on the substrate, but usually only one milliped at a time. The volume consumed is small and difficult to measure or weigh since it cannot be fed from a food dish. *Glomeris* are unlikely to climb into a dish to feed—if they did, they would likely dehydrate and die in the dish before the keeper could rescue them.

In some ways the reported details of a close relative, *G. balcanica,* reared in captivity is mirrored by those seen in wild and first-generation *G. pustulata* (McMonigle 2016). Both species are slow growing and probably take at least two years to mature. *Glomeris balcanica* reared in captivity require two and a half years to mature. They lay eggs in spring to mid-summer and molt late summer to early fall. According to observations of wild *G. balcanica* specimens, males undergo seventeen molts and females nineteen (Iatrou and Stamou 1989). Adults continue to molt after reaching sexual maturity, but at a slower rate. Adults of both species molt in late summer, though one of

Glomeris pustulata

the four *G. pustulata* adults molted again in February 2016 when kept at 70-74° F (21-23° C). (Iatrou and Stamou likewise maintained their laboratory specimens at 21-23° C.) Egg laying may not be controlled by the season. While *G. balcanica* laid eggs in spring to midsummer and molted late summer to early fall (Iatrou and Stamou 1990), *G. pustulata* laid eggs in late fall and winter in captivity (2015-2016). Iatrou and Stamou reported high mortality for the early growth stages of *G. balcanica* (nearly 90% combined attrition in the egg, protonymph, and first instar stages). Mortality was not observed for *G. pustulata*.

High mortality was reported for the early instars of *G. balcanica,* but no deaths were noted in the later stages of growth (Iatrou and Stamou 1990). *Glomeris pustulata* adult specimens molt in late summer (August to September), while small specimens molt more often.

Of six individuals acquired in 2015, one was immature, and another died after just two months. Unfortunately, the gender of the remaining four adults was uncertain. Male and female *Glomeris* should be very easy to identify because, like most millipeds, the males possess notable secondary sex organs (gonopods). However, sexing is not easy in practice. Attempting to pry the specimens open to identify the gender will potentially damage the animal and can result in a slow or rapid death. Alternatively, specimens may be placed on a clear surface and observed after they open on their own. This method is far safer, but it requires a certain measure of patience and available time. Sexing did not seem important because the few available reports suggested temperate *Glomeris* only lay eggs in the spring. The original specimens arrived in midsummer, which seemed to make gender determination immaterial. Nevertheless, a single tiny larva was observed in the adult's enclosure the last week of December 2015. Although this creature was incredibly small and pale gray in color (and the accidental introduction of a single foreign isopod immature was possible) the body segments appeared decidedly diplopodan.

Newly emerged larvae are quite small, though they are not terribly small relative to the adult size. Adults are only about twelve times as long. Larvae hatch out at just over 1 mm in length (2/3 that in diameter). By early March the largest offspring were no longer pale silver but had splotches of dark color. By early May of 2016 the largest had grown to 4 mm in length and were dark gray with pale yellowish spots in the same arrangement of the adult spotting. Something like adult pigmentation was still a few molts away.

In this first group, not a single egg was observed, despite searches through the substrate. They might have all hatched before

Glomeris pustulata molting

Glomeris aurita; Monte Arera, Italy
J. P. Oeyen (CC BY 4.0)

Glomeris connexa; Bavaria, Germany
Kryp (CC 0)

the search, but the egg capsules are sturdy. They appeared to have been laid individually because the variably sized immatures suggested staggered hatching events. The first larva showed up a month before the rest and molted twice before any others were observed. There were at least three separate waves of immatures, so they may have hatched mostly on different days. I am not certain if larvae were the product of more than one female. The largest number of individuals counted on the surface at one time was thirty-one, but from the results of limited digging, it seemed the total count was probably, at most, twice that many. With giant round millipedes (spirostreptidans and spirobolidans) there are often ten to thirty times as many animals buried as are ever seen on the surface at one time, but glomeridans live closer to the surface.

Providing food for very tiny millipeds can be difficult. They cannot travel far and can be easily damaged from handling. I provided food for the growing immatures by adding brown-sugar water and powdered milk alternately. A small pinch is cast over an area of the substrate and then misted so it soaks into the media. Freshwater algae showed some feeding, while dead leaves did not. A thin slice of fresh pumpkin was surrounded by five immature specimens the day after it was placed on the substrate. Iatrou and Stamou (1990) offered leaf litter and humus without any supplemental food. Lack of supplemental foods can greatly reduce survival rates for other types of millipedes (McMonigle 2012b).

The *G. pustulata* survival rate in the early instars seems to have been 100% from December 2015 to May 2016. However, egg survival was not observable, and shed exoskeletons could possibly have been confused with carcasses of small, pale larvae. If the survival rate was as high as it seemed, it may have been due to greater hardiness of this species (vs. *G. balcanica*) or a result of the access and types of food provided.

It is entirely possible that *G. balcanica* and *G. pustulata* differ in hardiness, egg laying times, and adult molts, but the differences are minor and represent a limited number of events. Immature survival in this report may be related to moss dirt offered as part of the diet. Iatrou and Stamou 1989 reported high mortality (around 50%) individually for the larva and first two free-living instars of *G. balcanica* (with a final survival of approximately 12%). Losses in the egg and larva were not recorded because they were not observed for *G. pustulata*, but no die-off was seen in free-living larva through 2022.

Hyleoglomeris sp. "Rubber Ducky"

The name for the species pictured here is an allusion to the popular rubber ducky isopod (which itself is an allusion to a bath-time toy). Specimens first showed up in 2021 in European holdings as imports from Thailand. Adults are around 0.5″ (12 mm) in length and the warning colors are spectacular. The color pattern is found in multiple related species and the collection location is uncertain, making species identification difficult. Like other glomeridans, they feed on standard millipede fares like dog food, melon, and rotten wood, but prefer the detritus under certain mosses.

Hyleoglomeris sp. "Yellow Striped"

Specimens appear to have black stripes on a yellow background, but the same question could be posed to the zebra. Either way, they are fantastically colored and easy to tell from the previous species and *Rhopalomeris* immatures by the pattern. This is another species that showed up on a vendor list in 2022 among a variety of imports from Thailand. This species is the smallest Old World glomeridan I have seen so far, at 0.3″ (<10 cm) unrolled. They are easy to care for and undergo post-ultimate molts in artificial conditions two or three times a year.

Onomeris sp. "Alabama"
The tiny North American glomeridans are almost never seen, or at least not noticed. They are very small, do not have a wide range, and do not occur in great numbers. Adult specimens are only about 0.1″ (3 mm) long and can easily be mistaken for an early instar offspring of the widespread, adventive isopod *Armadillidium vulgare*. Even for the milliped zoologist it can be difficult to confirm the identity without magnification because the tiny legs and body segmentation are at the lower end of human vision. The *Onomeris* species shown here has mated and produced small numbers of young in artificial habitats. Long-term survival and breeding seems to be related to substrate composition and moss detritus as with other glomeridans. It is difficult to observe supplemental food use, because of the small size, though they seem to ignore fruit and algae. Even with a group of dozens, it is difficult to gauge frass consumption to see if the substrate is adequate.

Candy Pill
Rhopalomeris carnifex (Pocock, 1887)
The standard coloration of *Rhopalomeris carnifex* of red and yellow with wide, black bands is known from Malaysia, Myanmar, and Thailand. Specimens of this color variety have often been sold under the name 'candy red' pill millipede. The 'candy yellow' look similar, but have reduced black stripes and no red. This less common coloration was named *Rhopalomeris carnifex* var. *pallida* (Pocock, 1887), given to specimens found by Pocock in the Mergui Archipelago (Myanmar). However, the specimens depicted here were collected on the mainland of Thailand, south near the city of Khanom, late 2018. Since the yellow form is not isolated to Mergui, and it has proven to be a normal variation, it is no longer considered a subspecies as first described. It was demoted to variety because museum specimens lack any subspecific characters (Golovatch et al. 2011). In addition to the pale-yellow form that lacks red, the common color variety can have reduced black banding, resulting in a very different look traded as 'candy corn.' Adult specimens vary in size and grow to plus or minus two centimeters when uncurled. Three or four slightly different color variations of the same themes

Onomeris sp.
Freshly molted animal next to exuvium

Hyleoglomeris sp. "Yellow Striped"
Thailand

Hyleoglomeris sp.
Thailand

Onomeris sp.
Tallapoosa Co., Alabama

(Clockwise from left) *Hyleoglomeris* sp. "Rubber Ducky"; *Hyleoglomeris* sp. "Yellow Striped"; *Glomeris pustulata; Rhopalomeris carnifex*

Rhopalomeris carnifex
2022 colors: "Candy Corn," "Tricolor," and "Dark"

Rhopalomeris carnifex "Dark"
(2022)

Rhopalomeris carnifex "Pallida"
(2019)

Rhopalmeris carnifex
2019 colors: "Pallida" and "Candy"

Rhopalomeris carnifex "Pallida"
Molting

Rhopalomeris sp.
Phetchaburi, Thailand
Luluchouette (CC BY 4.0)

showed up in 2022 Thai imports and some of these almost reached 1″ (2.4 cm).

The genus was originally described as *Glomeris,* but *carnifex* was transferred to *Rhopalomeris* in 1906. Like other glomeridans, they produce small droplets of clear, sticky fluid along the middle of the back when disturbed. This is often not exuded during handling and the production is limited. After a specimen uses the fluid, it can be weeks before it will do it again. Defensive fluid is odorless and has little to no effect on human skin.

Adult males are smaller than females. Adults molt two or three times a year, so a sexually mature specimen is not necessarily fully grown. Wild *R. carnifex* adult females have laid eggs which have developed into medium sized immatures in artificial enclosures. There has been some success in producing young from two color varieties (narrow and wide black bands), so the question is just when or if a consecutive generation will be successful. *Rhopalomeris* first showed up a few years ago as imports (McMonigle 2019). Mature specimens imported in 2019 lived as long as twenty-four months under artificial conditions.

High humidity seems to be best for this species, as they are usually kept in damp terrariums with minimal ventilation. The temperature should be maintained between 75-85° F (24-29° C).

Candy pill millipeds feed on rotten wood, dead leaves, fruit, fish flakes and pellets, crustose lichen, and mosses. However, they usually only consume substrate materials in quantity. These include compost, well-decayed and crumbled wood, and particulate debris accumulated beneath mosses. (The same moss species that grow without accumulating debris, on rocks and trees or on sandy soil, are usually ignored.) It is easy to overfeed processed fish food and fruit which mold quickly and are often ignored or only eaten in tiny quantities.

Trachysphaecra sp.
Limoges, France
Idvn (CC BY 4.0)

Order Julida

This order is closely related to the spirostreptidans, but the side segments of the lower part of the mouth (gnaothochilarium) meet towards the base and the front legs of the male are usually modified with hooks. The male's front legs are difficult to see on living specimens, even the rare big ones. Most species are tiny, 0.4-0.8″ (1-2 cm) in length and less than 1 mm in diameter. Previously the largest specimens I maintained were four ~3 cm *Bollmaniulus* sp. specimens from San Diego that lived from 2004 to 2010. However, the paeromopodidae, restricted to the west coast of the United States, can reach up to 6.5″ (16.5 cm). There was a photograph of a large *Paeromopus* in the 2012 *Millipedes in Captivity,* but I had never seen one and only first acquired *Californiulus* from a vendor in 2022.

Californiulus yosemitensis (Chamberlin, 1941)
This species is found primarily in California in the Sierra Nevada Mountain range, but as the most widespread member of the genus, occurs as far as southern Oregon and western Nevada. It is easy to discern from other *Californiulus* by color because the yellow band along the back has a solid, dark stripe down the middle. (Other species have no yellow band, or a yellow band with no line, or just a faint pattern down the middle.) Adults normally only grow to about 4.3″ (11 cm). They hide under pieces of bark in the enclosure and rarely, if ever, burrow in the substrate, except when they dig down to scrape out an underground cell for molting. Koi pellets and dog food are eaten enthusiastically, melon rinds and raw carrots are also eaten, but decayed wood and leaves are mostly ignored. Below 70° F (21° C), feeding ceases, which is unusual for temperate species. Like most creatures from the Pacific Northwest, temperatures greater than 80° F (27° C) may kill them. If handled gently, specimens are unlikely to exude defensive fluids.

Paeromopus angusticeps (Wood, 1864)
This large species from northern California is similar to *C. yosemitensis,* but it is a solid dark brown color. While it is supposed to grow as long as 5.9″ (15 cm), the mature female specimen I acquired was less than 4″ (9 cm).

Bollmaniulus montanae
Flathead Co., Montana

Uroblaniulus sp.
LaGrange, Georgia
Koby Yen (CC BY 4.0)

Californiulus yosemitensis

Paeromopus angusticeps
California

Ommatoiulus moreletii
Portugal

Blunt-Tailed Snake Milliped, *Cylindroiulus punctatus;* Berlin, Germany
Alexis_Orion (CC BY 4.0)

Spotted Snake Milliped, *Blaniulus guttulatus;* Versailles, France
Yves Bas (CC BY 4.0)

Striped Milliped, *Ommatoiulus sabulosus;* Denmark
Johan Kjær Prehn (CC BY 4.0)

White-Legged Snake Milliped, *Tachypodoiulus niger;* Merseyside, England
Steve J. McWilliam (CC0)

Order Platydesmida

Platy means 'flat,' but members of this order are commonly known as feather millipedes because of the long, widely spaced paranota or 'wings' on the side of each diplosegment. The paranota house a large part of the defensive glands (ozadenes). Defensive fluids from this order repel insects, but do not seem to effect human skin, nor offer a notable smell. Adult males guard the eggs and groups stay together in loosely structured colonies of varying stages. These social millipedes often congregate together in a perennial spiral or pinwheel pattern in nature (Wong 2017).

Brachycybe lecontii Wood, 1864
This feather milliped is widespread and common across the northeastern United States and is commonly maintained in artificial habitats. Adults are small, typically less than 1 ″ (2-2.5 cm), but the wide paranota give them some presence and clusters of specimens can be relatively large. *Brachycybe* can make a good colonial display, but they are difficult to keep long-term and sensitive to high temperatures. They seem to do best if kept at 65-72° F (18-22° C). Bracket fungi are the only known food, but they do not eat very much. Large, thick bracket fungi from the sides of big, old trees die and grow moldy when kept in high humidity and limited ventilation. *Brachycybe* dry out easily and would be very difficult to keep alive with good ventilation. Smaller, thinner types of bracket fungi that grow on branches, sometimes trunks, also appear to die, but do not necessarily grow moldy and can be fed on for up to six months.

Brachycybe producta Loomis, 1936
This pink feather hails from the West Coast and is easy to tell from other *Brachycybe* because of tiny notches on the back side of each paranota. Specimens tend to be a duller pink than *B. rosea,* and commonly have a dark line running down the back. They are also larger than *B. rosea*, a little over 2 cm. The dark line and coloration varies with feeding and growth. Like other *Brachycybe*, the natural habitat is cool, damp woods and it does best on the low side of room temperature.

Brachycybe rosea Murray, 1877
This Californian endemic is usually a beautiful hot pink to fluorescent orange color. Adults tend to be small, around 0.7 ″ (1.8 cm). They are found on rotting logs, often beneath bark. Groups rest together on flat surfaces where there are gaps between the wood and substrate. Like all the platydesmids, they live in habitats that neither get very hot, nor very cold, and are further insulated by the damp logs and fungus they live around.

Gosodesmus claremontus (Chamberlin, 1922)
The only member of its genus, this is one of the longest West Coast platydesmidans, commonly growing to over 1 ″ (2.8 cm). However, it has a much thinner body form and relatively short paranota. The elongate adults are a fraction of the width and mass of *Brachycybe*. It is traded as the 'pink boa,' as it resembles a tiny version of the clothing accessory, though to me it looks more purple than pink. Whereas *Brachycybe* normally rest against flat surfaces and are easy to observe in the artificial habitat, *Gosodesmus* burrow into wood and soil and can be difficult to locate even in a small enclosure.

Brachycybe leconteii

Brachycybe producta

Brachycybe rosea

Gosodesmus claremontus

Andrognathus corticarius, Montgomery Co., Virginia
P. L. Shorter, D. A. Hennen, and P. E. Marek. 2018. Cryptic diversity in *Andrognathus corticarius* Cope, 1869 and description of a new *Andrognathus* species from New Mexico (Diplopoda, Platydesmida, Andrognathidae). *ZooKeys* 786: 19-41. (CC BY 4.0)

Brachycybe guarding eggs
Sierra Nevada range, California
Sam McNally (CC BY-SA 4.0)

Brachycybe producta
Marin Co., California
Scott Loarie (CC0)

Platydesmida (*Fioria* or *Plutodesmus*)
Greece
Sarah Faulwetter (CC0)

Brachycybe rosea
Placer Co., California
Matt Berger (CC BY 4.0)

GLOSSARY

CARDO—basal mandibular segment

CLYPEUS—the lower edge of the milliped face just above the labrum

COLLUM—enlarged, dorsal plate above and behind the head, first segment of the milliped body

COMMENSAL—an organism that lives on another organism but does not help or harm the host; means "to eat at the same table"

DETRITIVORE—an animal that feeds on tiny fragments of dead plants and animals

DIPLOSEGMENT—milliped body segment formed by two fused segments

EGG CAPSULE—a hollow, generally spherical structure formed out of frass, leaf, or wood particles and containing a single egg; formed by many juliformia, pentazonia, polydesmida, and others

EXUDATE—defensive chemical fluids or gases secreted through the ozopores

EXUVIUM—shed exoskeleton

FRASS—the solid waste of millipeds

FRONS—front of the milliped head

GNATHAL LOBE—apical segment of the mandibles outfitted with biting and grinding structures

GNATHOCHILARIUM—bottom of the mouth formed by the maxillae

GONOPODS—secondary male sexual organs developed from one or both pairs of legs on the sixth or seventh segment, common to most milliped orders

HAPLOSEGMENT—any of the first four body divisions behind the head without paired organs

HOOD—dorsally visible male milliped sexual characteristic of expanded or enlarged segments between the head and gonopod segment; commonly widest at the seventh segment

HYPOPROCT—small, ventral anal scale; segment beneath the paraprocts

INSTAR—standard developmental stages of an immature arthropod terminating in a molt and prior to sexual maturity

LABRUM—upper jaw, just below the clypeus

L1, L2, ETC.—used internationally as the abbreviation for invertebrate larval instars. The *L* originates from the German term for instar, *Larven-stadium*. For example, a milliped that has molted three times after the protonymph stage is L4 (also 4th instar).

MARGIN—border or edge

OCELLUS (*pl.* OCELLI)—light detection organs

OZADENE—gland which produces defensive fluids

OZOPORES—tiny channels through which the defensive chemical is secreted, usually located on every body segment; repugnatorial pores

Paraprocts—anal valves

Paranota—lateral expansions of the body segments often very notable on polydesmidans and platydesmidans

Parthenogenesis—reproduction through the development of unfertilized eggs

Pleurite—body wall segment between the tergite and sternite

Post-Ultimate Molt—exoskeletal replacement after an arthropod is sexually mature

Preanal Ring—terminal body segment attached to the paraprocts and hypoproct

Protonymph—developmental stage between egg and first instar. This stage has no working legs, is not ambulatory, and the body segmentation is minimally defined; pupoid. Some authors split the protonymph into two or more stages.

Quinone—one of various yellow, organic acid compounds found in a wide variety of plants (and the milliped superorder Juliformia) that can be irritating to the eyes and mucus membranes

Sclerotize—to cause rigidity of the exoskeleton through addition of the protein sclerotin or substance other than chitin

Spermathecae—female receptacles for semen storage

Spiracle—external opening of the tracheal respiratory system; breathing hole

Stadium (*pl.* Stadia)—developmental stages including protonymph, instar, and adult, but not the egg stage. (Example: The post-ultimate female *Narceus americanus* is in its twelfth stadia.)

Sternite—ventral body segment where the legs are attached

Stipes—mandibular segment between the cardo and gnathal lobe

Telson—terminal segment or tail

Tergite—dorsal body segment; in most milliped orders it is fused to the pleurite and the juncture is obscured

Tömösvary Organ—advanced sensory pit located above the antennae which may serve multiple functions including for sound, pressure, odor, and humidity detection; found on millipeds, centipedes, and some other 'primitive' arthropods

Toxifabricant—chemical or chemicals synthesized by an organism for use in defense from predation

Trachea—internal breathing tube that opens at a spiracle and branches out to the body cells

Vulvae (*sing.* Vulva)—female sperm receptacles located behind the second pair of legs

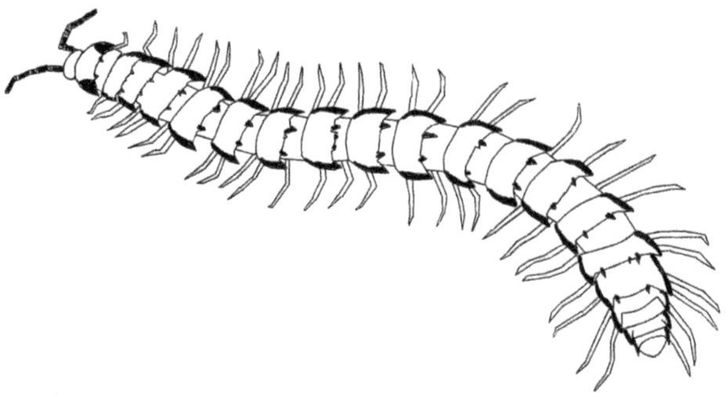

RESOURCES

Adkison, Perry. 2002-2008. Personal communications concerning millipeds and collection.

Bowman, Jared. 2000. Personal communication, *Narceus americanus* collection.

Carl, J. 1932. Diplopoden aus Süd-Indien und Ceylon. I. Teil. Polydesmoidea. *Revue Suisse de Zoologie* 39(17): 411-529.

Causey, Nell. 1957. *Floridobolus,* a new milliped genus (Spirobolidae). *Proceedings of the Biological Society of Washington.* 70(Dec. 31): 205-208.

Causey, Nell. 1959. *Narceus woodruffi,* new species, a Florida milliped (Spirobolida: Spirobolidae). *The Florida Entomologist* 42(3): 135-137.

Cook, O. F., and H. F. Loomis. 1928. Millipeds of the Order Colobognatha, with descriptions of six new genera and type species, from Arizona and California. *Proceedings of the United States National Museum* 72(18): 1-26.

Eisner, T. 2003. *For Love of Insects.* Cambridge, MA: Harvard University Press.

Enghoff, Henrik. 2005. The millipedes of Thailand (Diplopoda). *Steenstrupia* 29(1): 87-103.

Forster R. R., and L. M. Forster. 1980. *Small Land Animals of New Zealand.* Revised Edition. Dunedin, NZ: John McIndoe Ltd.

Golovatch, Sergei I. 1998. On several new or poorly-known Oriental Paradoxosomatidae (Diplopoda: Polydesmida), VI. *Arthropoda Selecta* 6(3-4): 35-46.

Golovatch, Sergei I. 2020. On several new or poorly-known Oriental Paradoxosomatidae (Diplopoda: Polydesmida), XXVIII. *Arthropoda Selecta.* 29(2): 161-172.

Golovatch, Sergei I., Jean-Paul Mauriès, Nesrine Akkari, Pavel Stoev, and Jean-Jacques Geoffrey. 2009. The millipede genus *Glomeris* Latreille, 1802 (Diplopoda, Glomerida, Glomeridae) in North Africa. *ZooKeys* 12: 47-86.

Golovatch, Sergei. I., R. L. Hoffman, and H. W. Chang. 2011. Identity of *Glomeris bicolor* Wood, 1865, and the status of the generic names *Hyleoglomeris* Verhoeff, 1910 and *Rhopalomeris* Verhoeff, 1906 (Diplopoda, Glomeridae). *Tropical Natural History* 11(1): 1-8.

Greenaway, Theresa. 2000. *Minipets: Centipedes and Millipedes.* Austin, TX: Steck-Vaughn Publishers.

Harrell, Wade. 2005. Congo black millipedes. *Invertebrates-Magazine* 5(4).

Hennen, Derek, and Jeff Brown. 2022. *Millipedes of Ohio Field Guide.* Publication 5527, Ohio Division of Wildlife.

Hoffman, Richard L. 1982. *Synopsis and Classification of Living Organisms.* New

York, NY: McGraw-Hill. (Diplopoda chapter, pp. 689-724.)

Hoffman, Richard L. 1999. *Checklist of the Millipeds of North and Middle America.* Special publication number 8, Virginia Museum of Natural History.

Hoffman, Richard L. 2006. Milliped or millipede? *Bulletin of the British Myriapod Group* 16: 1-2.

Hoffman, Richard L. 2008. Milliped toxifabricants. *Invertebrates-Magazine* 8(1).

Hopkin, Stephen, and Helen Read. 1992. *The Biology of Millipedes.* New York, NY: Oxford University Press.

Howarth, F. G., and W. P. Mull. 1992. *Hawaiian Insects and Their Kin.* Honolulu, HI: University of Hawaii Press.

Iatrou, G. D., and G. P. Stamou. 1989. Growth patterns of *Glomeris balcanica* (Diplopoda Glomeridae). *Proceedings of the 7th International Conference of Myriapodology.* pp. 217-221.

Iatrou, G. D., and G. P. Stamou. 1990. Studies on the life cycle of *Glomeris balcanica* (Diplopoda Glomeridae) under laboratory conditions. *Pedobiologia* 34(3): 173-181.

Likhitrakarn, N., S. I. Golovatch, and S. Panha. 2016. Review of the Southeast Asian millipede genus *Antheromorpha* Jeekel, 1968 (Diplopoda, Polydesmida, Paradoxosomatidae). *ZooKeys* 571: 21-57.

Locket, N. A. 1986. Albinism and eye structure in an Australian scorpion, *Urodacus yaschenkoi* (Scorpiones, Scorpionidae). *Journal of Arachnology* 14(1): 101-115.

McGavin, George C. 2000. *Insect Spiders and Other Terrestrial Arthropods.* New York, NY: Dorling Kindersley Inc.

McMonigle, Orin. 2002. Pink diplopodans. *Invertebrates-Magazine* 2(1).

McMonigle, Orin. 2004. AGBs. *Invertebrates-Magazine* 4(1).

McMonigle, Orin. 2005. *Giant Millipedes: The Enthusiast's Handbook.* Brunswick, Ohio: Elytra and Antenna.

McMonigle, Orin. 2005. "Glossy black pinkleg millipede" *Dendrostreptus macracanthus* (Attems). *Invertebrates-Magazine* 9(3).

McMonigle, Orin. 2006. Growing fire millipedes. *Invertebrates-Magazine* 6(3).

McMonigle, Orin. 2007. Featured invertebrate: "Thai rainbow" millipede, *Apeuthes* sp. *Invertebrates-Magazine* 6(3): 18.

McMonigle, Orin. 2008. Albino millipedes. *Invertebrates-Magazine* 7(3): 6-8.

McMonigle, Orin. 2008b. *Chicobolus spinigerus* (Wood). *Invertebrates-Magazine* 7(4).

McMonigle, Orin. 2009. *Centrobolus* sp. "Mozambique" "crimson blackleg millipede." *Invertebrates-Magazine* 9(1): 16-17.

McMonigle, Orin. 2010-2011. Cracking the polydesmid code. Parts 1 and 2. *Invertebrates-Magazine* 10(1) and 10(2).

McMonigle, Orin. 2012. Reproductive and developmental differences in the flameleg and rusty millipedes, Order Spirobolida, Family Pachybolidae, Subfamily Trigoniulinae. *Invertebrates-Magazine* 11(3): 9-11.

McMonigle, Orin. 2012b. *Millipeds in Captivity: Diplopodan Husbandry and Reproductive Biology.* Landisville, PA: Coachwhip Publications.

McMonigle, Orin. 2014. New Report for albinism in the Class Diplopoda (the millipeds): A North American representative of the Order Julida. *Invertebrates-Magazine* 14(1): 4-6.

McMonigle, Orin. 2015. *Glomeris pustulata* (Fabricius, 1781): A spotted pill milliped. *Invertebrates-Magazine* 15(1): 14-17.

McMonigle, Orin. 2016. Captive reproduction of *Glomeris pustulata:* A colorful European pill milliped, Part II. *Invertebrates-Magazine* 15(3): 2-5.

McMonigle, Orin. 2017. *Acladocricus* sp. (Spirobolida: Rhinocricidae) Ex. Negros Island, Philippines: Generations of breeding the giant blue. *Invertebrates-Magazine* 16(1): 7-9.

McMonigle, Orin. 2018. Featured invertebrate: Harlequin pill millipeds *Glomeris klugii* Brandt, 1833, in terraria. *Invertebrates-Magazine* 17(1): 15-18.

McMonigle, Orin. 2018b. Captive reproduction and longevity of *Apeuthes* sp. (Diplopoda: Spirobolida: Trigoniulidae) ex. Thailand: Two decades and counting. *Invertebrates-Magazine* 17(2): 8-11.

McMonigle, Orin. 2019. Color varieties and zoology of *Rhopalomeris carnifex* pill millipeds (Glomeridae: Doderiidae). *Invertebrates-Magazine* 18(2): 14-17.

McMonigle, Orin. 2020. Cracking the glomerid code. *Invertebrates-Magazine* 19(2): 9-14.

McMonigle, Orin. 2021. Similar asymmetric segmental anomalies in very different types of arthropods: Review and teratology documentation for *Dynastes tityus* (Scarabaeidae: Dynastinae), *Gromphadorhina oblongonota* (Blaberidae: Oxyhaloinae), *Lucihormetica subcincta* (Blaberidae: Blaberinae), *Heteropteryx dilitata* (Heteropterygidae), and *Porcellio expansus* (Porcellionidae). *Invertebrates-Magazine.* 20(2): 2-10.

McMonigle, Orin. 2021b. Featured invertebrate: Yellow striped milliped *Strophobolus immigrans?* (Chamberlain, 1920). *Invertebrates-Magazine.* 20(2): 11-12.

McMonigle, Orin. 2021c. Arthropod teratology conjoined segment update. *Invertebrates-Magazine* 21(1): 8-9.

McMonigle, Orin. 2022. Cultivation and longevity of a large northwestern U.S. round milliped *Tylobolus uncigerus* (Wood, 1864). *Invertebrates-Magazine* 21(2): 7-10.

McMonigle, Orin. 2022b. Teratology IV. *Invertebrates-Magazine* 22(1): 10-11.

Neal L. Evenhuis, David J. Preston, Keith T. Arakaki, and Clyde T. Imada. 2016. *Terrestrial Arthropod Survey of Halona Valley, Joint Base Pearl Harbor-Hickam, Naval Magazine Lualualei Annex, August-December 2015.* Contribution No. 2016-001 to the Hawaii Biological Survey.

Oba, Yuichi, M. A. Branham, and T. Fukatsu. 2011. The terrestrial bioluminescent animals of Japan. *Zoological Science* 28(11): 771-789.

O'Toole, Christopher. 1990. *Insects and Spiders.* New York, NY: Facts on File, Inc.

Ovalle, David. 2002. Millipedes move in, and the monkeys go wild. *Miami (FL) Herald* (August 8): 1A.

Pimvichai, Piyatida, Henrick Enghoff, Somsak Panha, and Thierry Backeljau. 2018. Morphological and mitochondrial DNA data reshuffle the taxonomy of the genera *Atopochetus* Attems, *Litostrophus* Chamberlin and *Tonkinbolus* Verhoeff (Diplopoda: Spirobolida: Pachybolidae), with descriptions of nine new species. *Invertebrate Systematics* 32(1): 159-195.

Preston-Mafham, Ken, and Rod Preston-Mafham. 2000. *The Natural World of Bugs and Insects.* London: PRC Publishing.

Putnam, L. G. 1958. Albinism in the migratory grasshopper, *Melanoplus bilituratus* (Wlk.). *Nature* 182: 1529.

Seeman, Owen, and Gerd Alberti. 2015. A new species of *Scissuralaelaps* (Acari: Mesostigmata: Laelapidae) from millipedes in the Philippines. *Systematic and Applied Acarology* 20(6): 707-720.

Shear, W. A. 2011. Class Diplopoda de Blainville in Gervais. 1844. pp. 59-164.

In: Zhang. Z.-Q. (ed.) Animal biodiversity: An Outline of Higher-level Classification and Survey of Taxonomic Richness. *Zootaxa* 3148: 1-237

Shelley, Rowland M. 1982. Confirmation of the status of *Euryurus mississippiensis* (Causey), with congeneric records from the Gulf States (Diplopoda: Polydesmida: Platyrhacidae). *Florida Entomologist* 65(2): 254-263.

Shelley, Rowland M. 1996. *The Biology of Millipedes,* book review. *Proceedings of the Entomological Society of Washington.* 98(1): 169-172.

Shelley, R. M., and G. B. Edwards. 2001. Introduction of the milliped, *Helicothomorpha holstii* (Pocock), in Florida (Polydesmida: Paradoxosomatidae). *Entomological News* 112(3).

Shelley, R. M., and S. D. Floyd. 2014. Expanded concept of the milliped family Spirobolidae (Diplopoda: Spirobolida, Spirobolidea): Proposals of Aztecolini n. tribe and Floridobolinae/ini and Tylobolini n. stats.; redescriptions of *Floridobolus* and *F. penneri,* both Causey, 1957, and *F. orini* n. sp.; Hypothesis on origins and affinities. *Insecta Mundi* 0357: 1-50.

Shelley, R. M., and S. I. Golovatch. 2011. Atlas of myriapod biogeography. I. Indigenous ordinal and supra-ordinal distributions in the Diplopoda: Perspectives on taxon origins and ages, and a hypothesis on the origin and early evolution of the class. *Insecta Mundi* 0158: 1-134.

Shelley, R. M., and P. T. Lehtinen. 1999. Diagnoses, synonymies, and occurrences of the pantropical millipeds, *Leptogoniulus sorornus* (Butler) and *Trigoniulus corallinus* (Gervais) (Spirobolida: Pachybolidae: Trigoniulinae). *Journal of Natural History* 33: 1379-1401.

Sierwald, P., and J. Spelda. 2020. Millibase. Accessed at http://www.millibase.org/ on 2020-10-22.

Sigling, Shura. 2010. *Professional Breeders Series: Millipedes.* Frankfurt: Edition Chimaira.

Stanek V. J. 1969. *The Pictorial Encyclopedia of Insects.* Middlesex England: Hamlyn Publishing Group.

Walls, Jerry G. 1999. *The Guide to Owning Millipedes and Centipedes.* Neptune City, NJ: T.F.H. Publications.

Werner, Floyd, and Carl Olsen. 1994. *Insects of the Southwest.* Tucson, AZ: Fisher Books.

Wong, Victoria. 2017. *Natural History of the Social Millipede Brachycybe lecontii (Wood, 1864).* Thesis submitted to Virginia Polytechnic Institute and State University.

Yoram Yerushalmi, Eran Tauber, and Meir Paul Pener. 2001. Phase polymorphism in *Locusta migratoria*: the relative effects of geographical strains and albinism on morphometrics. *Physiological Entomology* 26(2): 95-105.

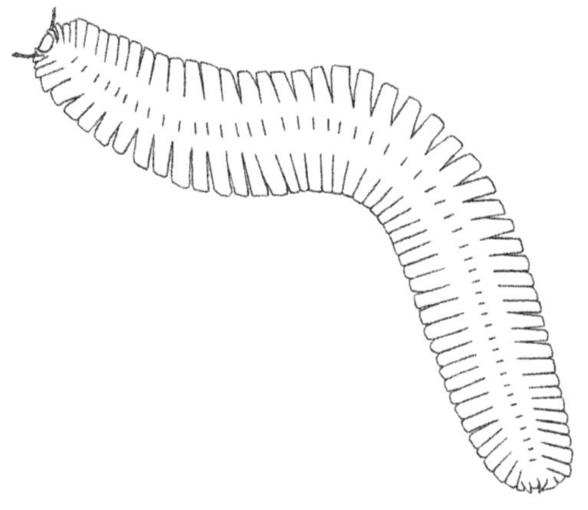

ASSASSIN BUGS,
WATERSCORPIONS,
AND OTHER HEMIPTERA
REPRODUCTIVE BIOLOGY AND
LABORATORY CULTURE METHODS
Orin McMonigle

CENTIPEDES IN
CAPTIVITY
The Reproductive Biology
and Husbandry of
Chilopoda
Orin McMonigle

BREEDING THE WORLD'S
LARGEST LIVING ARACHNID
AMBLYPYGID BIOLOGY, NATURAL HISTORY, AND CAPTIVE HUSBANDRY
ORIN MCMONIGLE

INVERTEBRATES
FOR EXHIBITION
ORIN MCMONIGLE
INSECTS, ARACHNIDS, AND
OTHER INVERTEBRATES SUITABLE
FOR DISPLAY IN CLASSROOMS,
MUSEUMS, AND INSECT ZOOS

MILLIPEDS IN CAPTIVITY
DIPLOPODAN HUSBANDRY AND
REPRODUCTIVE BIOLOGY
ORIN A. MCMONIGLE

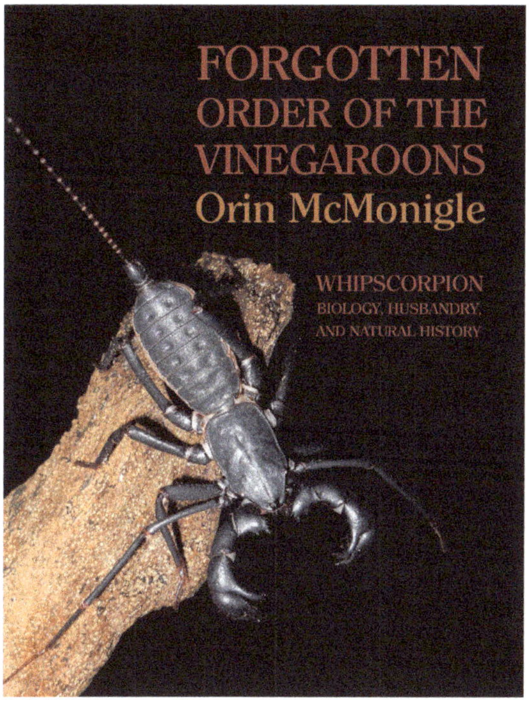
FORGOTTEN
ORDER OF THE
VINEGAROONS
Orin McMonigle
WHIPSCORPION
BIOLOGY, HUSBANDRY,
AND NATURAL HISTORY

PILLBUGS AND
OTHER ISOPODS
CULTIVATING VIVARIUM CLEAN-UP CREWS AND
FEEDERS FOR DART FROGS, ARACHNIDS, AND INSECTS
Orin McMonigle

KEEPING THE
PRAYING MANTIS
Orin A. McMonigle

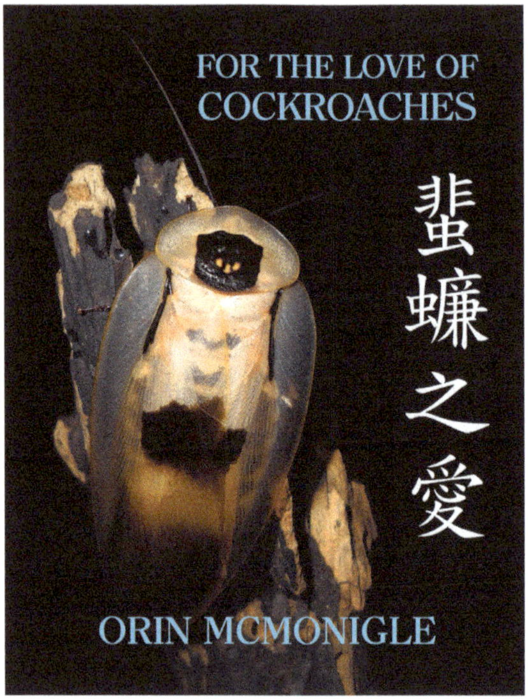
FOR THE LOVE OF
COCKROACHES
蜚蠊之愛
ORIN MCMONIGLE

THE ULTIMATE GUIDE
TO BREEDING BEETLES
COLEOPTERA LABORATORY CULTURE METHODS
ORIN MCMONIGLE

BREEDING THE VAMPIRE
AND OTHER CRABS
(BRACHYURA AND ANOMURA IN CAPTIVITY)
Husbandry, Reproduction,
Biology, and Diversity
ORIN MCMONIGLE

ORIN MCMONIGLE
ISOPOD
ZOOLOGY
Biology, Husbandry,
Species, and Cultivars

www.ingramcontent.com/pod-product-compliance
Ingram Content Group UK Ltd.
Pitfield, Milton Keynes, MK11 3LW, UK
UKHW060221061026
15625UKWH00007B/426

9 781616 465506